Understanding DNA and Gene Cloning

A GUIDE FOR THE CURIOUS

Understanding DNA and Gene Cloning

A GUIDE FOR THE CURIOUS

Third Edition

KARL DRLICA

Public Health Research Institute
455 First Avenue
New York, NY

John Wiley & Sons, Inc.

New York • Chichester • Brisbane • Toronto • Singapore • Weinheim

Acquisitions Editor	David Harris
Executive Marketing Manager	Catherine Faduska
Production Editor	Deborah Herbert
Designer	Kevin Murphy
Cover Designer	Lynn Rogan
Manufacturing Manager	Mark Cirillo
Illustration	Rosa Bryant

This book was set in 10 on 12 Palatino by TCSystems, Inc. and printed and bound by Courier Westford. The cover was printed by Phoenix Color Corp.

Recognizing the importance of preserving what has been written, it is a policy of John Wiley & Sons, Inc. to have books of enduring value published in the United States printed on acid-free paper, and we exert our best efforts to that end.

The following material was adapted from K. Drlica, DOUBLE-EDGED SWORD: THE PROMISES AND RISKS OF THE GENETIC REVOLUTION, © 1994 by Karl A. Drlica and reprinted by permission of Addison-Wesley Longman Publishing Co., Inc.: descriptions of patterns of inheritance, Mendelian inheritance, DNA fingerprinting, and Figures 13-2, 13-3, 13-4, 14-3, and 14-5.

Library of Congress Cataloging-in-Publication Data
Drlica, Karl.
    Understanding DNA and gene cloning: a guide for the curious/
Karl Drlica.
        p.   cm.
    Second ed. published in 1992.
    Includes bibliographical references and index.
    ISBN 0-471-13774-X (pbk. : alk. paper)
        1. Molecular cloning.   2. Recombinant DNA.   3. Genetic
engineering.   I. Title.
QH442.2.D75   1996
574.87'3282—dc20   96-23077
                                                                                          CIP

Printed in the United States of America

10 9 8 7 6 5 4 3 2 1

To Ilene, for many years of support

PREFACE

An explosion of knowledge is shaking the science of biology, an explosion that will soon touch each of us. At its center is chemical information—information that our cells use, store, and pass on to subsequent generations. With this new understanding of life comes the ability to manipulate chemical information. We can restructure the molecules that program living cells. Already this new technology is being used to solve problems in such diverse areas as waste disposal, synthesis of drugs, treatment of cancer, plant breeding, and diagnosis of human diseases. The new biology is also telling us how the chemicals in our bodies function; we may soon be programming ourselves and writing our own biological future. When this happens, each of us will be confronted with a new set of personal and political choices. Some of these difficult and controversial decisions are already upon us, and the choices will not get easier. Informed decisions require an understanding of molecular biology and recombinant DNA technology; this book is intended to provide that understanding.

Molecular biology is a science of complex ideas supported by test tube experiments with molecules. Consequently, the science has remained largely inaccessible to those without a knowledge of chemistry. I have tried to change that situation—this book requires the reader to have little or no background in chemistry. Consequently, our discussion of DNA starts at a much more elementary level than is commonly found in publications such as *Scientific American* and the *New York Times*. Chemical processes and molecular structures are described by means of analogies using terms familiar to nonscientists. Technical terms are kept to a minimum; where they must be introduced, they are accompanied by definitions. These definitions are grouped, often in expanded form, in the glossary. To provide a feeling for informational molecules, I have also introduced a few details about how they are manipulated experimentally. Integration of these details should help remove the mystery from gene cloning and expose the elegance and simplicity of the technology.

This is the third edition of *Understanding DNA*. When I completed the first edition, gene cloning was largely an academic subject. By the time the second edition was finished, information about gene cloning had become commercially important: stock brokers, patent attorneys, and judges were making decisions that required knowledge of DNA and its activities. The first edition had even been cited as a reference for DNA patent cases by the U.S. Court of Appeals. Now there has been another quantum leap in molecular genetics, this time into the realm of health prediction. Today everyone in the industrialized world needs to understand the implications of gene cloning to take advantage of the new technology and to minimize the risks associated with release of personal genetic information. Thus *Understanding DNA* had to be broadened to be useful for personal decision making. That meant adding chapters dealing with human genetics. When the new material had been drafted, however it became obvious that the book should be restructured. The present version has four parts: basic molecular biology, manipulation of DNA, insights gained through the use of gene cloning, and human genetics; Part III contains a chapter on retroviruses, since they are so important. Readers interested in a more extensive, but elementary discussion of the benefits and risks of the genetic revolution are referred to my other book, *Double-Edged Sword* (Addison-Wesley, 1994).

Originally, *Understanding DNA* was written for college students, with readers ranging from nonscience majors to potential biology majors. While the audience now extends to the general public, several elements have been retained to keep *Understanding DNA* appropriate for classroom use. Among these are "Questions for Discussion" at the end of each chapter. Some of these questions have specific answers to reinforce a point; others are open-ended to stimulate additional reading. Many of the questions introduce information that would dilute the main themes if included in the primary text. Another teaching aid is the glossary. Vocabulary is a key aspect of learning molecular biology, and the reader should expect to refer frequently to the definitions listed. As an instructor I found that administering simple glossary quizzes early in a course overcomes some of the vocabulary barriers. A third aid is the list of additional readings, which has also been expanded. Most of the new entries are from *Scientific American* because the articles are of high quality, they are at the appropriate level when used in conjunction with *Understanding DNA*, and *Scientific American* is readily available.

I thank the following persons for helping with the third edition: David Betsch, David Bourgaize, Shirley Chapin, Marila Gennaro, Sherban Iordanescu, James Moulds, Ellen Murphy, Harry Ostrer, George Sideras, Richard Sinden, and Todd Steck. I also thank the staff at John Wiley & Sons, including Sally Cheney, Catherine Donovan, David Harris, Deborah Herbert, and Brenda Griffing for encouragement and skillful production of the work.

Karl Drlica

INTRODUCTION

Media coverage of the O.J. Simpson murder trial has made DNA a household word, but few people fully appreciate how important DNA science will be to their own lives. A hint of the growing importance can be gleaned from casual reading of major newspapers. One example involves a baby who would have been plagued with infections due to a faulty immune system. The condition had been discovered by prenatal DNA testing, however, and when the baby was born, doctors saved cells from the umbilical cord. The cells were then engineered with DNA to correct the immune system defect and placed back in the baby. This experiment may allow the baby to live a normal life.

A second example concerns DNA analyses that are gradually entering the military through a program in which blood and tissue samples are routinely obtained from recruits for identification purposes, in the event that war-related activities result in dismemberment. Fearing an invasion of genetic privacy, two marines chose court martial over participation in the program.

In a third example, some of the families that participated in identifying a "breast cancer gene" were outraged at geneticists for withholding information about the breast cancer status of subjects' children. The scientists argued that a child is incapable of giving the informed consent that physicians require before they reveal such devastating news. The parents, however, wanted the information to help guide and hopefully protect their children.

Still other cases involve rape. For example, the attorneys of men wrongly convicted of rape have had DNA analyses performed on the semen-stained clothing that had been used as evidence. In two dozen cases, it was demonstrated that the semen in question could not have come from the defendant; thus innocent men were released from prision. A British approach to tracking down rapists has involved thousands of men. Sometimes all the males in a community are asked to volunteer for DNA tests to flush out the perpetrator. For these and each of the preceding examples, the key point is that information in DNA can be used to track people and their diseases.

Before long, the medical profession will be able to look into our health future through a wide variety of genetic tests. Individuals will be identified who are especially susceptible to particular ailments, including problems arising from specific environmental dangers. A person who happens to be in a susceptible group will be advised to arrange his or her lifestyle to avoid the particular danger. Thus people who get skin cancer quickly when exposed to sunlight reduce their risk by shading themselves and by staying indoors much of the time. Those who have a high probability of contracting breast or colon cancer monitor their bodies more carefully than most people and sometimes opt for preemptive surgery. Still others are in danger of lethal reactions to anesthetics commonly used during surgery. By foreknowledge, their doctors can avoid the danger. Even scientists are taking advantage of their own work. A woman working on a gene involved with heart disease tested her own blood and found that she was predisposed to the disease. She immediately started taking the vitamin that helps reduce the problem. With time, the list of predictable, treatable problems will grow to the point that most of us can benefit from genetic testing.

On the downside, many of the maladies that can be identified currently have no cure. This poses a dilemma because people will not necessarily want to know that they are destined to suffer from an incurable disease decades from now. Doctors will not be able to tell them what to do. Even when a physician's advice can help us make a clear-cut decision, such assistance may be difficult to obtain because so many tests will be available. For example, proper counseling requires about 30 minutes per disease; for a hundred diseases we would need 3000 minutes. It will not be possible for each of us to obtain 50 hours of technical consultation from our doctors—we will be forced to use our own understanding of DNA and genetics to make decisions.

Genetic discrimination will amplify our need to be informed. In the past, families carrying genetic ailments such as Huntington disease were considered to be cursed. The uncontrollable shaking made it easy to identify the afflicted, who were sometimes incarcerated or put to death as witches. In modern times, it's usually careers and financial status that suffer as members of Huntington disease families are deemed uninsurable or are rejected from lengthy professional training programs. In the near future, improvements in the detection of many different genetic blemishes will reveal that each of us is predisposed to some condition we would prefer to avoid. As this knowledge reaches our employers and insurance carriers, they may be tempted to reduce their costs by cutting us from their rolls. Even when discrimination is illegal, we will have to learn to recognize bias when it can harm us and take appropriate action.

Managed health care brings an additional dimension to the discrimination dilemma: for some of us, our doctors are employed by our insurance company or by our employers. Press reports revealing that doctors are sometimes pressured by companies to reveal medical records of employees have fed worries about breaches in doctor–patient confidentiality. The message is quite explicit: for some types of test, such as that for HIV infection and those for genetic disease, avoid your company doctor.

Just the fear of discrimination will make us cautious about what is written in our medical records, since we know that this material is not secure (life insurance companies often require us to release medical histories prior to issuing coverage). The fear will not be limited to ourselves, since genetic diseases are passed from one generation to the next. Thus we will worry that our disclosure will lead to problems for our children and grandchildren. And if we don't tell our doctors everything about our family medical history, we take on the burden of self-diagnosis. We will each have to learn medical genetics.

In the future, it will become clear to everyone that DNA information can be changed in individuals, in families, and in nations. At that point we will be confronted by a new set of decisions. At the individual level, gene therapy protocols are already adding good genes to body cells that have a bad one. Should this expensive treatment be made available to everyone through government subsidies? At the family level, genetic tracking can follow hereditary diseases as they pass from one generation to the next, enabling geneticists to single out persons harboring genetic disease and assist them with family planning. Often, afflicted fetuses can be identified early in pregnancy. However, the definition of "afflicted" is controversial, and there is disagreement over when it should be applied in individual abortion decisions. Since selective abortion could in principle eradicate a disease from a family, governments may be tempted to initiate genetic improvement programs as they seek ways to reduce health care costs. Here the decisions become quite profound because we as a society would be taking a small step toward altering the very nature of humankind.

As the Age of the Gene Hunter comes into full bloom, each of us will have the opportunity to get much more out of our bodies. At the same time, we will face a variety of risks, stemming largely from the ability of other people to know about our prospects for health. Difficult decisions lie ahead, decisions that require an understanding of DNA.

K. D.

CONTENTS

C H A P T E R   O N E
PREVIEW
Life as Interacting Molecules

Overview

Information governing the characteristics of all organisms is stored in long, thin molecules of deoxyribonucleic acid (DNA). DNA molecules contain regions (genes) that specify the structure of other molecules called proteins. Protein molecules in turn control cellular chemistry and contribute to cell structure. Biologists can now obtain large amounts of specific regions of DNA. The general strategy involves first cutting the DNA into small fragments. The fragments are then moved into single-celled organisms, often bacteria or yeast. Conditions are set up so the fragments become a permanent part of the microorganism, and pure cultures of bacteria or yeast containing a particular DNA fragment are then obtained. The fragments are removed and used for further study, for detecting disease, for changing the cellular chemistry of another organism, and for producing large quantities of specific proteins of medical or industrial value. Since microorganisms receiving DNA fragments are being changed in unknown ways, gene cloning experiments were originally viewed as *potentially* dangerous. Regulations were established to minimize the potential health hazards. Many types of recombinant DNA have now been constructed and studied; no harmful effects have been observed, and almost all biologists now consider the cloning of genes to be safe. Use of the cloned genes raises a variety of other issues ranging

1

from the production of herbicide-resistant plants to genetic modification of humans.

INTRODUCTION

In a general sense biologists have solved the riddle of **heredity,** the question of how offspring come to resemble their parents. The answer lies in the chemical behavior of submicroscopic structures called molecules. (All boldface words are defined in the glossary.) At the center of this new understanding is a giant molecule called **deoxyribonucleic acid (DNA).** This book is about DNA, the chemical that specifies features such as eye color and blood type. DNA in our cells influences all our physical characteristics, as is true for every living **organism** on earth. This book is also about **genetic engineering, recombinant DNA,** and **gene cloning.** In particular, it is about how gene cloning works and about some of the things we have learned from doing it. The first chapter provides a framework; a discussion of DNA structure and function is started, and a few comments are made about fundamental features of **atoms, molecules, enzymes,** and **cells.** Gene cloning is introduced to provide a sense of direction, followed by brief discussion of safety issues. The chapter concludes by outlining the types of information presented in the rest of the book.

The basic unit of life is the cell, an organized set of chemical reactions bounded by a **membrane** and capable of self-perpetuation. Our bodies are collections of trillions of cells working together, with each cell having its own identity and function. For example, liver cells cluster to form livers, and skin cells attach to each other to cover our bodies. With few exceptions, every cell contains all the information required for an independent existence; indeed, under the right conditions human cells can be removed from the body and grown in laboratory dishes.

The information necessary to control the chemistry of the cell (i.e., the chemistry of life) is stored in the long, thin fiber called DNA. DNA fibers are found in every cell except mature red blood cells, and they dictate how a particular cell behaves. Thus, DNA controls our body chemistry by controlling the chemistry of each of our cells.

Isolated DNA looks like a tangled mass of string (Figure 1-1). Our cells, which are generally less than a **millimeter** long, contain about 2 meters of DNA specially packaged to fit inside. DNA can be bent, wrapped, looped, twisted, and even tied in knots. Many DNA molecules are circles, which are sometimes found interlinked like a magician's rings. Thus DNA is very flexible, at least in terms of three-dimensional structure. But in terms of information content, DNA is quite rigid, for the same information must pass from generation to generation with little change.

One of the goals of this book is to explain how information is stored in DNA, how it is reproduced, and how it is used. For now, the important concept is that distinct regions of DNA contain distinct bits of information. The specific regions of information are called **genes.** In some ways DNA is similar to motion picture film. Like film, DNA is subdivided into *frames* that make sense when seen in the correct order. In DNA the *frames* correspond to the letters in the genetic code, which are described in Chapter 2. When a number of frames or genetic letters are organized into a specific combination, they create a scene in the case of film and a gene in the case of DNA (Figure 1-2).

Information in genes is used primarily for the manufacture of **proteins,** the chainlike molecules that fold in a precise way to form specific structures. Some proteins contribute to the architecture of the cell, while others directly control cell chemistry. Occasionally we can easily see the effects of particular genes and proteins; for example, a small group of genes is responsible for determining eye color. It is the specific information in the DNA, in the genes, that makes human beings different from honey bees or fir trees.

ATOMS AND MOLECULES

Each DNA fiber is a molecule, a group of atoms joined together to form a distinct unit. Several points are important for understanding discussions of atoms and molecules. First, all forms of matter are composed of submicroscopic particles called atoms. Second,

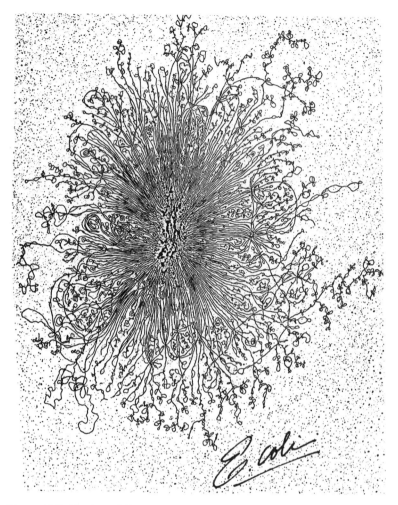

Figure 1-1 Electron Micrograph of a DNA Molecule Released from a Bacterium. The long, threadlike material is DNA, which is about one milli-meter long or about 1000 times the length of the bacterium from which it was taken. The molecular details of how this DNA is compacted and packaged to fit in the cell are not yet understood. The electron micrograph is of a purified, surface-spread *E. coli* chromosome prepared by Ruth Kavenoff and Brian Bowen. The line under the *E. coli* signature represents 2.5 micrometers. (Copyright © with all rights reserved by DesignerGenes Posters, Postcards, T-shirts, etc., P.O. Box 100, Del Mar, CA 92014; ttnx82a@ prodigy. com.)

4

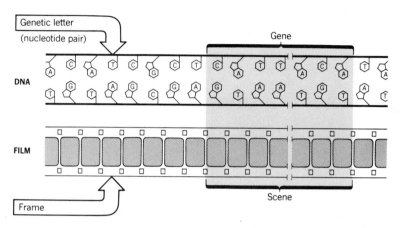

Figure 1-2 Comparison of DNA and Motion Picture Film. The frames of movie film correspond to genetic letters (**nucleotides**) in DNA. When properly organized, the frames form a scene in film, and the genetic letters form a gene in DNA. DNA contains many genes, and each one stores information affecting a chemical process that occurs in living cells. Genes are generally hundreds to thousands of nucleotide pairs long. For illustrative purposes, only a part of a gene is shown in the figure; slashes drawn through the DNA and the film indicate the omission of many nucleotides and frames.

molecules are very specific combinations of atoms, and the combinations have distinctive properties. Third, atoms and molecules can join with each other or with single atoms to form new molecules—that is, new combinations of atoms whose properties differ from those of the starting materials. Such interactions are called **chemical reactions.** For example, hydrogen gas is two atoms of hydrogen, and oxygen gas is two atoms of oxygen. If you combine an oxygen atom with two hydrogen atoms, you get water, a molecule very different from either oxygen or hydrogen. Fourth, the number of **elements,** or atoms of different kinds, is small (about 100). The number of types found in living material is still smaller (the major ones are listed, along with their common abbreviations, in Figure 1-3). The small number greatly simplifies the process of converting one substance to another. A fifth point

Name	Symbol	Atomic weight	Properties of pure element
Hydrogen	H	1.01	Light, colorless gas
Carbon	C	12.01	Hard solid (diamond, graphite)
Nitrogen	N	14.01	Colorless gas
Oxygen	O	16.00	Colorless gas
Fluorine	F	19.00	Pale greenish gas
Sodium	Na	23.00	Reactive silver metal
Magnesium	Mg	24.31	Light, silvery metal
Phosphorus	P	30.97	White, red, or yellow nonmetal
Sulfur	S	32.06	Yellow solid
Chlorine	Cl	35.45	Yellow-green gas
Potassium	K	39.10	Light, silver-white metal
Calcium	Ca	40.08	Soft, silvery metal
Manganese	Mn	54.94	Hard, brittle metal
Iron	Fe	55.85	Silvery grey metal
Copper	Cu	63.54	Malleable reddish metal
Zinc	Zn	65.37	Blueish-white metal
Selenium	Se	78.96	Red or grey semimetal
Molybdenum	Mo	95.94	Tough, silvery metal
Iodine	I	126.90	Violet crystalline nonmetal

Figure 1-3 Atoms (Elements) Commonly Found in Living Material. Like all matter, living organisms are composed of specific combinations of atoms. The chemical symbol and the relative size (atomic weight) are listed for each of the elements (i.e., atom types) found in organisms. (Adapted from E. O. Wilson et al., *Life*, Sinauer Associates, Sunderland, MA.)

is that large molecules such as DNA are composed of many smaller groups of atoms. By understanding the properties of the smaller groups, it is possible to predict the behavior of large molecules formed when multiple groups are joined together. Thus it is not necessary to know the precise position of every atom in DNA to understand how the giant molecule acts.

To discuss the properties of molecules, it is often useful to draw pictures in which the relative positions of the component atoms are specified. These pictures, or structural formulas, help explain the properties of molecules. Frequently the pictures are abbreviated, and often some of the atoms are omitted for clarity: compare the different ways of representing sugar structure illustrated in Figure 1-4.

When atoms form a molecule, they are held together by forces called chemical bonds. Chemical bonding is generally discussed in terms of the structure of atoms. Atoms are composed of a nucleus surrounded by a cloud of electrons that circle the nucleus much as the planets orbit the sun. Atoms differ in nuclear size and electron number; the number and arrangement of electrons determine bonding patterns. Thus bonding patterns are characteristic of the specific atoms involved in an interaction. For example, carbon almost always forms four bonds to other atoms, and hydrogen forms only one (see Figure 1-4*a*). Some types of chemical bond are very strong, such as those formed when two atoms share an electron (such bonds are represented by the lines between the atoms in Figure 1-4). The strong bonds keep the atoms of each molecule together; they make each molecule a discrete physical entity. Other bonds are much weaker. The attractive forces that result from atoms having different electrical charges, for example, are easily broken. However, if many weak bonds form between two molecules, they can collectively be quite strong. Weak bonds are often responsible for bringing distant regions of a large molecule close or for holding two different molecules together (for an example, skip ahead to Figure 2-5*a:* the dotted lines represent weak bonds). For two molecules to be held together, the molecules usually have to fit tightly. The weak bonds allow biological molecules to recognize each other, to come together, and to separate when conditions change. Thus one can think of DNA as a molecule that contains information specified by the arrangement of its atoms, which are held together by strong bonds. Weak bonds provide the forces that hold the two DNA strands together and allow DNA to interact physically with other molecules.

ENZYMES AND CHEMICAL REACTIONS

In the chemical reactions of life, molecules are broken down into simpler molecules, built up into more complicated molecules, or

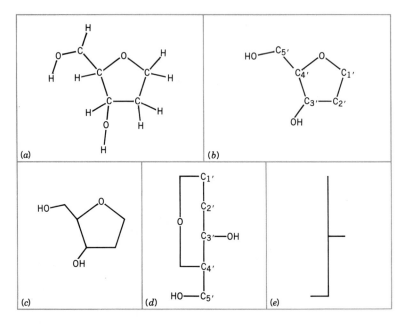

Figure 1-4 Symbolic Representations of a Molecule. The arrangement of atoms in a molecule can be illustrated in a number of ways. Frequently atoms are deleted from a diagram for clarity. Sometimes little attention is paid to the precise spatial orientation of the atoms (bond angles and lengths) because such details would obscure the point of the drawing. Here the arrangement of atoms in the sugar called deoxyribose is depicted in four ways. None of these includes information about spatial orientation. (a) All the atoms are shown; the lines between atoms represent chemical bonds that hold the atoms together. (b) The hydrogens, except for the two attached to oxygen, have been removed from the diagram for clarity. (c) The structure is the same as in (b), but the carbon atoms are not explicitly indicated. (d) The carbon–oxygen ring in (b) is redrawn in a way that is useful for showing directionality in DNA, a point that is developed in later chapters. The carbon atoms are numbered in (b) and (d) to help relate the two diagrams. (e) The atoms have been eliminated from the structure in (d) in a way that emphasizes the two hydroxyl (OH) groups.

simply rearranged to create slightly different molecules. Over the years chemists have been able to describe many of the rules that govern these reactions. In some ways, organizing a series of chemical reactions is similar to tailoring a shirt: the cloth is cut into specific shapes, the front and back are joined together, the sleeves and collar are stitched, and then they are attached to the body. Finally buttons are added. Each step is analogous to one of the chemical reactions in the series; through a series of steps, the original piece of cloth is converted into a new form.

Although one person working alone can make a shirt, the analogy to cell chemistry is clearer if we imagine each task to be the assignment of a different specialist—a cutter, a stitcher, and so on. If shirts are to be produced efficiently, the tasks must be performed in a particular order. So it is with the work of **enzymes,** the specialized protein molecules that control the chemistry of cells. The key idea is that the chemicals in every cell in every organism react in an orderly fashion; the enzymes provide that order. Some enzymes are very valuable engineering tools, particularly those that cut DNA and those that join DNA molecules together.

Most enzymes are proteins (a few RNA molecules have been discovered that act as enzymes; they are called **ribozymes,** and they are briefly discussed in Chapter 11). Proteins are long, linear molecules that are much like beaded necklaces folded in a specific way (Figure 1-5). Some proteins contain hundreds or even thousands of beads. Each bead is called an **amino acid.** There are 20 common types of amino acid, so an astronomical number of ways exist in which a chain can be put together. Consequently, it is easy to show that there could be many, many different types of protein. The chemical properties of the amino acids in a protein determine precisely how the protein folds and thus how it acts.

Enzymes are catalysts. They speed up chemical reactions without themselves being consumed in the reaction. Many enzymes do this by binding to specific molecules in such a way that bonds between certain atoms in the molecules can be formed or broken more readily than in the absence of the enzyme. In many cases the molecules involved in the reaction fit into pockets or clefts present on the enzyme surfaces. There the molecules are often distorted to make them more reactive. It is important to realize that the properties of a particular

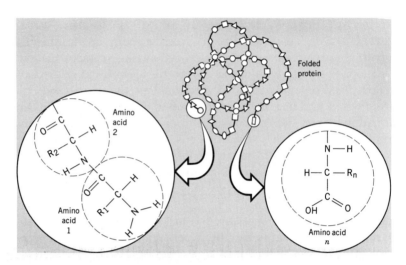

Figure 1-5 Protein Structure. Proteins are long chainlike molecules composed of many amino acids. The 20 common amino acid types are distinguished by the arrangement of atoms in the group labeled **R** in each amino acid. The precise folding of the protein chain is determined by the arrangement of R groups. Each protein has chemically distinct ends. The amino acid labeled **1** is called the amino-terminal amino acid. Note that its nitrogen has one more hydrogen than that of amino acid **2,** and its nitrogen is bonded to only one carbon atom. Amino acid **n** is the carboxy-terminal amino acid. At this end of the protein the terminal carbon is bonded to two oxygen atoms. The N-C bond joining two amino acids is called a **peptide bond.**

enzyme—that is, how it controls a particular chemical reaction—are determined by the *order* of the amino acids in the chain. Consequently, knowing how the cell specifies the precise order of amino acids in a protein is central to understanding how enzymes are made. Such knowledge is the key to understanding heredity, for the characteristics of enzymes and other proteins make a cell what it is. The information determining the order of amino acids in every protein is stored in DNA. We discuss the conversion of information in DNA into protein (Figure 1-6) in more detail in Chapter 4.

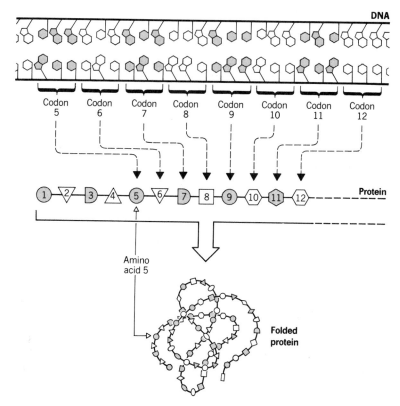

Figure 1-6 Relationship of Information in DNA to Protein Structure.
Information in DNA is arranged in a series of 3-letter words called **codons.**
Each codon specifies a particular amino acid. For example, codon 5 has
information for the fifth amino acid in the protein. The overall shape and
activity of a protein depend on the precise order of amino acids. The information for this order is stored in the DNA. The two DNA strands are wound
around each other as discussed in Chapter 2 (see Figure 2-1); for clarity,
winding of the DNA strands is not shown here. Also omitted for simplicity
are messenger RNA and other aspects of the conversion of information in
DNA into protein (see Chapter 4).

CELLS AND CHROMOSOMES

As pointed out above, the molecules of life assemble themselves into reproducing units called cells. The thin membrane that keeps the ensemble of molecules together is a double layer of fatty molecules containing an assortment of proteins that form channels and pumps to move molecules in and out of cells. Inside cells DNA is packaged into structures called **chromosomes.** Frequently organisms are divided into two general categories on the basis of how DNA is packaged. The more primitive types, which include **bacteria,** are called **prokaryotes.** Their DNA intimately contacts the rest of the cell. More advanced organisms, including ourselves, are called **eukaryotes.** Our chromosomes are grouped together inside a membrane-bound bag called a nucleus. This strategy of separating cellular components from each other by enclosing them in a membrane is common to eukaryotes. The resulting structures are called **organelles. Mitochondria** are organelles that contain the proteins that convert the chemical energy of sugar into a more useful form called ATP. **Chloroplasts** trap energy from sunlight and convert it into chemical energy by forming sugars, and organelles called **lysosomes** serve as garbage disposals, where complex molecules are broken into forms that are easily recycled.

Most cell types reproduce by dividing in half. The newborn cells then grow larger, duplicate their DNA, and undergo division themselves. This series of events, called the cell cycle, is repeated over and over as microorganisms reproduce or as bodies develop from fertilized eggs. Near the end of the eukaryotic cell cycle, the chromosomes condense into structures that are easily seen with a **light microscope.** By that point the chromosomes have duplicated to form pairs that remain attached as they assemble at the center of the cell. The nuclear membrane has broken down, and a member of each duplicated pair moves to one end (pole) of the cell while the other member of each pair moves to the opposite pole. A membrane forms between the two sets of chromosomes and creates two cells. At about the same time, a nuclear membrane forms around each set of chromosomes. The chromosomes then become more loosely packed, making the information in DNA accessible to guide formation of additional cellular components.

As multicellular organisms develop, cells specialize to form partic-ular tissues and organs. With few exceptions all cells in a particular individual organism have the same information in their DNA mole-cules. Different cell types arise from use of information from different regions (genes) of DNA. As adulthood is reached, most of the cells stop dividing, although many retain the capacity to grow again if needed to heal a wound. Some cells are even programmed to die so an organ can form and function properly. In all cases the programs are carried by DNA molecules.

Bacterial cells lead much simpler lives. Many types just grow and divide. They are much smaller than eukaryotic cells, and they gener-ally have only one chromosome, which does not condense when cells divide. However, the biochemical principles that operate in bacteria are the same as those running us. Thus bacterial systems can be used as simple examples to illustrate many of the principles of life.

GENE CLONING

Gene cloning is the process of obtaining many copies of a short, specific piece of DNA through reproduction of a microorganism containing the DNA piece. Operationally, gene cloning consists of performing a series of biochemical manipulations analogous to baking a cake: recipes are available for each step. In general, to clone a region of DNA one must first cut the DNA into specific pieces. The cutting tools usually produce DNA fragments of so many different types that individual types cannot be separated in a straightforward way. Instead, one-celled microorganisms are used to carry out the separation. Billions of microorganisms are mixed with the fragmented DNA in such a way that the DNA moves inside the target organisms and takes up permanent resi-dence there. Conditions are adjusted to ensure that no microorgan-ism receives more than one piece of DNA. The microorganisms are then scattered onto a solid surface. There they grow and reproduce; each forms a small cluster. The biologist then tests the clusters to determine which one contains the DNA fragment of interest. That cluster of cells is saved and allowed to reproduce many more times. Finally the cells are broken open, and the DNA fragment of interest is removed.

Cloned genes can be used to change the characteristics of organisms; thus we can expect gene cloning to have a major impact on agriculture as new breeds of plants and animals are developed. In the area of medicine, genetic testing is giving some couples new reproductive options by helping them identify fetuses destined to become very ill children. Testing is also enabling adults to discover whether they are particularly prone to cancer, and gene therapy has shown promise as a type of treatment for a rare childhood disease.

Cloned DNA can also be used to produce medically important proteins. Generally a protein that controls a chemical reaction is very specific to that reaction. Consequently, if a person produces a defective protein, it is sometimes possible to inject a replacement protein without interfering directly with other chemical reactions of the body. For example, some diabetics fail to produce sufficient quantities of the protein called **insulin** and are unable to properly control their **sugar metabolism.** Daily injections of insulin can often control the disease. Before the development of gene cloning, insulin could be obtained only by an expensive process of extracting the protein from hog pancreas. Now human insulin genes have been placed in bacteria. Here the genes are expressed; that is, insulin is made inside bacteria. Large quantities of insulin are produced by bacteria, and it is easier to obtain insulin from bacteria than from pancreas tissue. Moreover, the engineered bacteria produce human insulin, an important feature for diabetics who have become allergic to hog insulin.

THE SAFETY CONTROVERSY

Shortly after the first gene cloning experiments were completed in the early 1970s, scientists realized that this type of biomanipulation might pose health hazards. No danger had been demonstrated, but one could imagine a number of scenarios frightening enough to make good science fiction copy. Suppose, for example, that the gene containing the information for botulism **toxin** were placed inside a harmless bacterium. If large numbers of this new, toxin-producing microorganism were accidentally released into the environment, a few might find their way into the digestive tracts of humans. There they would multiply, for the human digestive tract is one of the normal habitats for the type of bacterium most commonly used in gene cloning experiments (but not for the organism that normally produces botulism toxin). Then, it was feared, anyone exposed to the *engineered* bacteria would soon die. Since little was known about the ecological relationships between common laboratory bacteria and man, there was no way to determine whether such a scenario could be realized. Thus it seemed prudent to use recombinant DNA technology with great care.

The scientists who developed gene cloning recognized the need for precautions. They tried to control the use of the **cloning vehicles,** the biological tools used to transfer genes, but it soon became apparent that they would be unable to do this by themselves. Gene cloning involves straightforward laboratory procedures, and the discoverers knew that soon hundreds of scientists all over the world would be conducting experiments with recombinant DNA. To be effective, the control system would require much more muscle than a handful of scientists could flex.

In the United States, a regulatory system was set up by the National Institutes of Health (NIH), the federal agency responsible for funding most recombinant DNA research. A set of guidelines was established that focused on containing recombinant organisms inside laboratories. In some cases the investigators could use standard laboratory techniques, whereas experiments of other types had to be conducted in specially constructed rooms isolated from the environment. A few experiments were simply disallowed. The responsibility for compliance rested on each institution involved in recombinant DNA research. At stake was the institution's access to federal research grants. Many thousands of experiments have been performed using the NIH guidelines, and fears based on earlier predictions of accidents with catastrophic consequences appear to be unfounded.

Most scientists have always viewed the risks as very small for several reasons. First, recombinant DNA, as an isolated material in a test tube, is not dangerous; only under carefully controlled conditions can it be transferred into living cells. Second, recombination (i.e., the forming of new combinations of DNA segments) occurs in nature and is not in itself dangerous. Third, bacteria do not often escape from laboratories and establish infections. Even highly evolved **pathogens** (disease-causing organisms) have been success-fully contained in laboratories, and isolated cases of laboratory infec-tion have not led to epidemics. This is partly because the strains of bacteria used for gene cloning do not survive well outside the laboratory; indeed, most of them are genetic cripples, often requiring special nutrients for growth. Fourth, the chance is small that one of these bacteria would be converted accidentally into a pathogen. To infect their hosts, pathogenic organisms rely on sophisticated mechanisms involving a number of different genes that have been carefully honed over millions of years of evolution. For all these reasons, prudent caution in the conduct of experiments with recom-binant DNA has been deemed adequate.

Another type of controversy has now emerged. Should genetically engineered organisms be deliberately released into the environment? The issue is whether engineered bacteria or plants, however useful, will displace other organisms from their normal ecological niches. And if so, what are the consequences, how long would we have to wait to see them, and could we reverse them?

PERSPECTIVE

Understanding gene cloning requires two types of knowledge: a grasp of the concepts of molecular biology and a familiarity with the necessary laboratory manipulations. Both aspects are discussed in the chapters that follow. It is important to realize that working with huge numbers of very small items requires strategies that may not initially be obvious. The molecular *scissors* used to cut DNA into small fragments make many, many cuts. Thousands of different pieces of DNA are produced. Locating the single, desired fragment of DNA and separating it from all the other pieces is a formidable task. Unfortunately, DNA cannot be run through an editing machine like a piece of film; genes in DNA are too small to be seen by the human eye. Even with the highest powered microscopes, DNA containing thousands of genes looks like a featureless piece of string (see Figure 1-1). To find a specific gene, cloners blindly separate the DNA fragments; then they examine the different fragment types biochemically until they locate the desired ones. The separation process involves putting individual DNA fragments *randomly* into many, many microbial cells.

Understanding how DNA itself is manipulated requires several types of information. First it is necessary to describe some of the basic ideas of molecular genetics. Sometimes called the central dogma, the key concept is that genetic information flows from DNA to RNA to protein. Chapter 2 presents the fundamentals of DNA structure, and Chapter 3 describes how DNA is reproduced (replicated). Since many of the protein tools used to engineer genes are components of DNA replication systems, the discussion of replication provides an opportunity to explain how biologists handle molecules. Chapter 4 then focuses on gene expression, the process of converting genetic information from a DNA to a protein form. Controlling the expression of a particular gene is often important for getting the gene to work properly. With Chapter 5 the emphasis shifts to manipulating DNA by introducing the topic of microbial growth. Chapter 6 describes plasmids and phages, the submicroscopic vehicles that carry DNA fragments into bacterial cells. Then cutting and joining DNA molecules are discussed (Chapter 7), followed by a consideration in Chapter 8 of the many ways complementary base pairing is used in recombinant DNA work. At that point the individual steps of gene cloning can be tied together into a single recipe (Chapter 9). Chapter 10 then deals with ways in which cloned genes are used to dissect life processes. The next two chapters move to a discussion of what we have learned from our ability to pull out small pieces of DNA and analyze them. Chapter 11 focuses on features that were not easily predicted by the central dogma, and Chapter 12 addresses three related topics relevant to everyone alive: retroviruses, AIDS, and cancer genes. By that point you will have a general understanding of one of the major strategies biologists are using to discover how living cells work. To allow you to use that information, the book concludes with two chapters on human genetics.

Questions for Discussion

1. The earth and all living things on it are composed of atoms. What distinguishes living objects from nonliving ones? The critical factor cannot be growth and movement, for inanimate things such as crystals grow, and both water and air move.

2. To work with molecules, one must have ways to detect them and to measure how many are present. To experience measuring something you cannot see, first imagine that you are blindfolded and sitting at a dinner table. Then devise a way to determine how much water is in your glass without sticking your finger in.

3. Certain elements can occur as radioactive isotopes, unstable forms in which the atoms spontaneously disintegrate. The disintegration, called radioactive decay, releases high energy particles and radiation characteristic of the particular type of atom. DNA and other biological molecules can be prepared in such a way that some of their atoms are radioactive. This measure greatly facilitates the detection of the biological molecule because radioactivity can be measured relatively easily. List some of the ways that radioactivity can be detected.

4. Among the elements that have radioactive forms are plutonium, hydrogen, carbon, iodine, cesium, and phosphorus. By referring to Figure 1-3, determine which of these can be incorporated into biological molecules.

5. Each type of radioactive atom decays at a characteristic rate. For example, half the atoms in a common form of radioactive phosphorus decay every 2 weeks. If you prepare DNA containing radioactive phosphorus and then take a 2-month holiday, what fraction of the radioactivity initially in the DNA will be present when you return to work?

6. We are composed of molecules. Most of our molecules are quite stable at body temperature, but within our bodies chemical reactions (conversions of one type of molecule to another) are occurring continuously. Obviously our bodies have ways to

make at least some of the molecules unstable without raising the temperature to very high levels. How might this occur?

7. Discuss the ways in which gene cloning might affect your life.

8. Calculate the number of ways in which 20 different types of amino acid can be arranged in a **peptide** (a chain of amino acids) 10 amino acids long.

PART ONE
BASIC MOLECULAR GENETICS

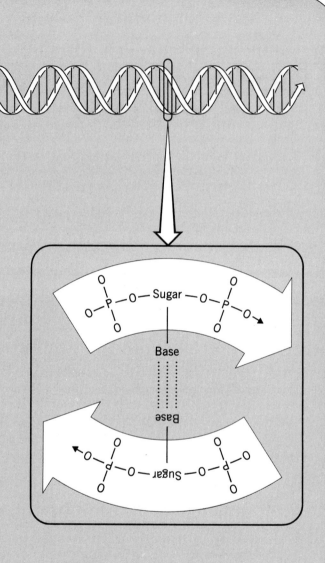

In DNA the strands run in opposite directions and are held together in part by interstrand interactions between the bases of the nucleotides.

STRUCTURE OF DNA
Two Long, Interwound Chains

Overview

DNA is a long, threadlike molecule composed of subunits called nucleotides; in one sense it resembles a two-stranded, beaded necklace. DNA molecules can be very long, sometimes containing more than a hundred million nucleotides. There are four different types of nucleotide (abbreviated A, T, G, and C), and it is through the specific order of the nucleotides that genetic information is stored in DNA. The nucleotides in the two strands pair with each other so that an **A** in one strand is always opposite a **T** in the other, and a **G** is always opposite a **C**. This rule allows the two strands to act as templates for the formation of new strands. The two ends of a DNA molecule differ in chemical reactivity: DNA has distinct left and right ends, and the two strands run in opposite directions.

Genes are discrete stretches of nucleotides that contain information specifying the sequence of amino acids in proteins. It takes three nucleotides to specify a particular amino acid; that is, specific nucleotide triplets or codons correspond to specific amino acids. Specific combinations of nucleotides also signal the beginning and end of a gene. In terms of information content, DNA is a very stable molecule, and the information is faithfully reproduced and passed from one generation to the next. But

23

the DNA molecule should not be considered to be static in terms of its three-dimensional structure, for it interacts with many cellular proteins that can dramatically alter its shape.

INTRODUCTION

To understand how DNA controls the activities of our cells and thus the activities of each of us, it is necessary to briefly consider the structure of DNA. One level of discussion focuses on the chemical structure of the subunits, the nucleotides. The structure of the nucleotides holds the key to understanding how DNA is able to reproduce and how information contained in DNA is utilized. Another level of structure, the order of the nucleotides, determines the information content; it is also the order that determines in part where specific proteins will bind and cause DNA to carry out its many activities. A procedure for determining the order of nucleotides is described in Chapter 10. A third aspect of structure deals with DNA as a whole, as a long, flexible molecule. Three-dimensional structure is important when considering how the cell gains access to information in specific regions of DNA because twisting and wrapping of DNA can greatly alter the ability of proteins to bind to it.

CHEMICAL STRUCTURE OF DNA

In the first chapter the organization of information in DNA was described by drawing parallels between this biomolecule and motion picture film. But to describe the chemistry of DNA, other analogies must be developed. One can think of DNA as a long, thin string composed of two strands wound around each other much like strands in a rope (Figure 2-1*a*). But closer inspection, using biochemical rather than microscopic methods, reveals that each strand is composed of tiny **subunits.** Thus, the DNA strands are more accurately thought of as two interwound strings of beads, with each

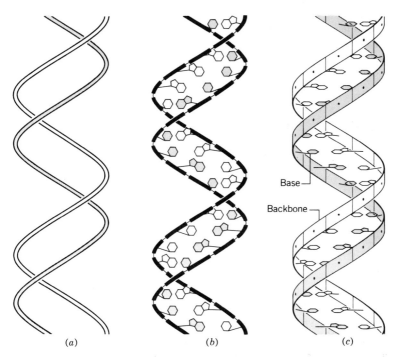

Base—

Backbone—

(a)   (b)   (c)

Figure 2-1 Schematic Representations of DNA. (a) Two interwound strands. **(b)** Two interwound beaded chains. **(c)** Double helix of two interwound strands with bases on the inside and backbone on the outside.

string containing millions of minute subunits linked together as suggested in Figure 2-1*b*. Chemists call the subunits nucleotides, and they have found that each nucleotide is composed of three parts (Figure 2-2): a flattened ring structure called a **base,** a **sugar** ring called deoxyribose, and a **phosphate.** Alternating sugars and phosphates form the backbone of the DNA. The bases are attached to the sugars and are located between the backbones of the DNA strands, lying perpendicular to the long axis of the strands (Figures 2-1*c* and 2-3). As the backbones of the two strands wind around each other, they form a double helix (Figures 2-1*c* and 2-3), leading to the popular expression for DNA.

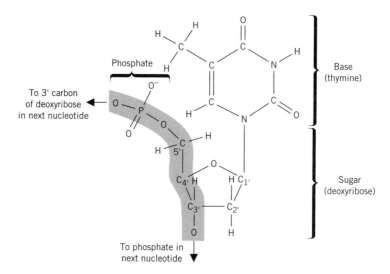

Figure 2-2 Structure of a Deoxyribonucleotide. A nucleotide is composed of three parts, the base (in this case it is thymine), the sugar deoxyribose (see also Figure 1-4), and a phosphate. One nucleotide is joined to the next by bonds between deoxyribose and phosphate. The carbon atoms of the deoxyribose are numbered 1′ through 5′. The 3′ carbon of the sugar attaches to a phosphate from the nucleotide immediately to its right in DNA. The phosphate bound to the 5′ carbon of the sugar attaches to the nucleotide at its left by binding to the 3′ carbon of the sugar in that nucleotide. The shaded region represents the atoms that are a part of the backbone of DNA.

The bases tend to stack one on top of another, much like steps in a spiral staircase. The bases come in four varieties, popularly abbreviated A, T, G, and C. The letters stand for **adenine, thymine, guanine,** and **cytosine,** the chemical names of the bases. Since each nucleotide contains only one base, the nucleotides can also be identified by the same four letters. These four nucleotides are precisely ordered in DNA, and it is through this arrangement of nucleotides that cells store information. The principle is similar to the Morse code, which transmits information in combinations of two symbols, dots and dashes.

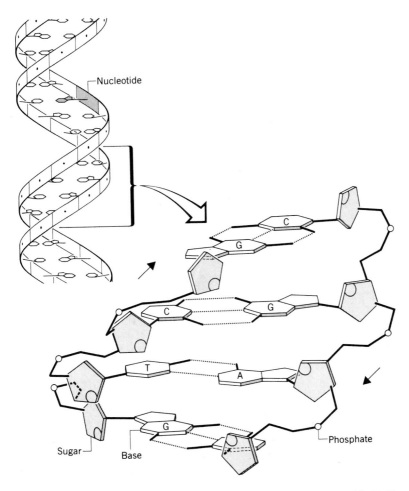

Figure 2-3 Schematic Diagram of a Short Section of a DNA Double Helix.
Each DNA strand is composed of chemical structures of three types: bases,
sugars, and phosphates. A nucleotide is a unit composed of one base, one
sugar, and one phosphate. The sugars and phosphates connect to form the
backbone of each strand, and a base attaches to each sugar. The four different
bases in DNA are represented by the letters A, T, G, and C. The bases of
one strand point inward and toward those of the other. Attractive forces
called **hydrogen bonds** (represented by broken lines) exist between the
bases of opposite strands and contribute to holding the two strands together.
The two strands run in opposite directions; notice how the sugars in one
strand seem to point upward while those in the other seem to point down-
ward. See Figure 2-4 for additional description of directionality.

27

Extensive examination of DNA has led to the identification of three rules that govern DNA structure. First, a single DNA strand does not have branches. Consequently, the information is stored in a simple line. Second, the ends of a DNA strand are chemically different. DNA has directionality much like a line of elephants connected trunk to tail. One end of the line is terminated by a trunk and the other by a tail. Each end is named according to the sugar carbon at that end (3' or 5'). By convention the **5' end** is generally drawn as the left end (Figure 2-4*a*); when two strands of opposite polarity are paired, the upper strand has its 5' end on the left (Figure 2-4*b*).

The third rule merits its own paragraph because it is the most important concept in molecular genetics. The rule stipulates that when two DNA strands come together and form a double helix, bases must fit together in a precise way. Whenever an A occurs in one strand, a T must occur opposite it in the other strand. Likewise, G always aligns opposite C. Corresponding bases in opposite strands are called **nucleotide pairs** or **base pairs.** Only when the bases are properly paired will two DNA strands fit together. This third principle is called the **complementary base-pairing rule.** It is important to note that the two strands of DNA are **complementary,** NOT identical; identical nucleotides do not form base pairs. One can imagine that the bases opposite each other fit together like electrical plugs and sockets (Figure 2-5): only the correct pairs coincide. As a result of the millions of tiny bases fitting together and stacking on top of each other, the two strands of DNA tend to stick tightly together. That makes the DNA double helix a stable structure; temperatures near that of boiling water are required to separate the strands.

ORGANIZATION OF INFORMATION IN DNA

As pointed out above, the subunits of the DNA strands, the nucleotides, are the chemical basis for storage of information in DNA. If we return to the film analogy introduced in Chapter 1 (Figure 1-2), we see that the units defined as nucleotides correspond to the frames

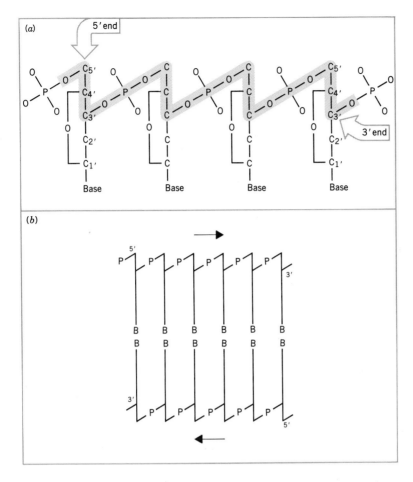

Figure 2-4 Nucleic Acid Directionality. (a) The ends of nucleic acids are distinct and have different chemical properties. For illustrative purposes, the sugar portion of the nucleic acid has been stretched out and only its carbon (C) and oxygen (O) atoms are shown. The sugars are joined by phosphates, and the backbone of the chain is illustrated by the shaded region. The five carbon atoms in each sugar group are assigned numbers 1′ through 5′. The ends of the nucleic acid are designated by the symbols 5′ and 3′. At the 5′ end of the chain a terminal phosphate (PO₄) is joined to sugar carbon number 5′, and at the 3′ end a terminal phosphate is connected to sugar carbon number 3′. A nucleic acid can end with either a phosphate group or a hydroxyl (OH) group; enzymes that act on the ends of a nucleic acid often prefer a particular type of end. (b) The two DNA strands have opposite polarity. Here, deoxyribose is reduced to a line (see Figure 1-4e), phosphate to a **P,** and base to a **B.** By convention, the upper strand has its 5′ end on the left.

(*a*) **Structural formulas**

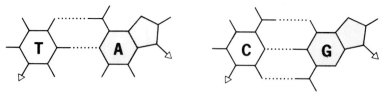

(*b*) **Prongs and sockets**

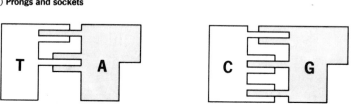

Figure 2-5 Complementary Base Pairing. (a) Structural formulas for thymine:adenine (T · A) and cytosine:guanine (C · G base pairs). The bases are flat structures composed of hydrogen, carbon, nitrogen, and oxygen atoms. The solid lines represent chemical bonds between these atoms. Arrows indicate points at which the bases attach to sugars. The dotted lines are hydrogen bonds, weak attractive forces between hydrogen and either nitrogen or oxygen. Notice that there are two hydrogen bonds between adenine and thymine, and three between guanine and cytosine. This difference in hydrogen bonding comprises the structural explanation for complementary base pairing. **(b)** A prongs-and-sockets analogy for base pairing. The hydrogen atoms in each hydrogen bond are represented as prongs, and the oxygen or nitrogen atoms are depicted as sockets. The attractive forces are weak; consequently, perfect fits are required for base pairing to occur.

in a motion picture film. That is, the genetic letters mentioned earlier represent the chemical base name abbreviations A, T, G, and C. Thus, information is stored in DNA by the specific sequence of a 4-letter code (A, T, G, C), which reads in a line along the DNA like frames in a film. In physical terms, a gene is a stretch of DNA ranging from hundreds to thousands of nucleotides; it corresponds to a scene in motion picture film.

We can begin to define a gene more precisely by considering features important for converting information from DNA into protein. First, a gene has a beginning and an end, signaled by short stretches of nucleotides. Second, information in a gene is arranged as words rather than as individual letters. This is because proteins, the linear, chainlike molecules made from information in DNA, have subunits (amino acids) of 20 types rather than the 4 subunits (nucleotides) used to store information in DNA. A quick calculation predicts how many nucleotide letters would be necessary to code for each amino acid, that is, how many letters in DNA correspond to each letter in a protein. Obviously one nucleotide in DNA cannot correspond to one amino acid in protein because there are five times more types of amino acid than types of nucleotide. Likewise, the nucleotides cannot be read in sets of two because 4 different nucleotides taken two at a time can produce only 16 possible pairs, 4 short of the minimum number. The code can be read in threes: 4 nucleotides taken three at a time give 64 possible triplets. This is more than enough to specify the 20 amino acids as well as the necessary punctuation, such as start and stop signals. Many experiments have confirmed the prediction that the genetic code is read as triplets of nucleotides.

Triplets in DNA that correspond to amino acids are called **codons.** When the code is read by the machinery of the cell, the information is first converted into an **RNA (ribonucleic acid)** form by synthesis of a new RNA molecule. RNA is similar to DNA but with the base **uracil** (U) substituted for the thymine (T) in DNA. The sugar in RNA is also slightly different from that in DNA. The genetic code, in its RNA form, is shown in Figure 2-6. The process of converting information from DNA to RNA is called **transcription,** which also refers to RNA synthesis. The RNA is called **messenger RNA.** The information in messenger RNA (mRNA) is next changed into amino acid sequences in protein by a process called **translation.** Both transcription and translation are described in Chapter 4.

There is no punctuation between the codons. Thus it is important that the **reading frame** of the nucleotide code be established correctly—the start signal must be in the right place. This principle can be illustrated by the following sentence read as 3-letter words:

First base	Second base: U	Second base: C	Second base: A	Second base: G
U	UUU, UUC } Phe UUA, UUG } Leu	UCU, UCC, UCA, UCG } Ser	UAU, UAC } Tyr UAA, UAG } TERM	UGU, UGC } Cys UGA TERM UGG Trp
C	CUU, CUC, CUA, CUG } Leu	CCU, CCC, CCA, CCG } Pro	AAU, CAC } His CAA, CAG } Gln	CGU, CGC, CGA, CGG } Arg
A	AUU, AUC, AUA } Ile AUG Met	ACU, GCC, ACA, ACG } Thr	AAU, AAC } Asn AAA, AAG } Lys	AGU, AGC } Ser AGA, AGG } Arg
G	GUU, GUC, GUA, GUG } Val	GCU, GCC, GCA, GCG } Ala	GAU, GAC } Asp GAA, GAG } Glu	GGU, GGC, GGA, GGG } Gly

Figure 2-6 The Genetic Code. Four nucleotides in RNA, taken three at a time, can form the 64 combinations shown: 61 correspond to amino acids in protein; the other 3 are stop (termination) codons. The amino acids are listed by the 3-letter abbreviations given in the **amino acid** entry in the glossary. Notice that all amino acids except tryptophan and methionine have more than one codon. Where there are multiple codons for amino acids, the base in the third position seems to have the least meaning. The bases in each codon are written with the 5′ end of the RNA on the left and the 3′ end on the right (see Figure 2-4 for description of 5′ and 3′ ends). In its DNA form, the code would contain Ts instead of Us. (Adapted from P. B. Weisz and R. N. Keogh, *The Science of Biology*, McGraw-Hill, New York.)

JOE SAW YOU WIN THE BET

If you read in sets of three but start out of register, at the letter O instead of J, you would end up with a meaningless string:

OES AWY OUW INT HEB ET

Consequently, whenever an engineer inserts genes into a new DNA molecule, the correct start signal must be present to establish the right reading frame, both for the amino acids being joined and for the stop signal.

DNA AS A LONG FIBER

Chromosomal DNA molecules are often more than a thousand times longer than the cell in which they reside. Thus DNA must be folded and compacted. One level of organization is the folding of DNA into large loops, each containing about 100,000 nucleotide pairs. Another level, found in higher organisms, is the wrapping of 200 nucleotide pair stretches of DNA around specific proteins called **histones.** This wrapping generates a series of ball-like structures along the chromosome. Some biologists think that a similar wrapping of DNA occurs in bacteria, but the ball-like structures are not readily detected with these cells.

Many activities of DNA involve pulling portions of the duplex apart, at least temporarily. Since the strands are wound around each other, strand separation requires the ends to rotate. But many DNA molecules are circles, lacking ends for rotation. Cells are able to overcome this problem because they contain **topoisomerases,** enzymes that can cut DNA, allow intact strands of DNA to pass between the cut ends, and then seal the ends back together (Figure 2-7). This type of activity can introduce and remove twists in DNA, tie and untie DNA knots, and even link or unlink circular DNA molecules. Thus DNA is a flexible, dynamic molecule that responds in a variety of ways to the action of proteins inside the cell.

ISOLATION OF DNA

To remove some of the mystery from the study of molecules, it is useful to consider how DNA can be isolated and handled. It turns out that DNA is so long and stringy that it can be easily separated from most other cellular components. A common practice is to first treat a tissue sample with detergent to break open the cells. Detergent also helps to strip proteins from the DNA. Sometimes salt is added

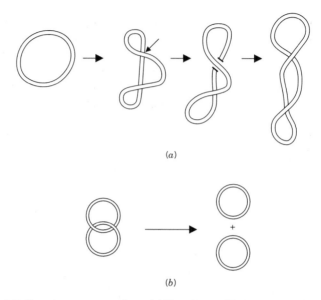

(a)

(b)

Figure 2-7 Topoisomerase action. (a) Topology of DNA supercoil introduction. A circular DNA molecule will occasionally become distorted, with one
region lying on another. A topoisomerase called DNA gyrase can perform
the equivalent of breaking the bottom strands (arrow) and passing the top
strands through the break. Sealing the break leaves the DNA twisted. (b)
Decatenation. When circular DNA is reproduced, it can end up as two
interlinked circles. Topoisomerases present in cells can separate the circles
by breaking one double helix and passing the other through.

to the sample, to induce some of the proteins to **precipitate** (clump).
Clumped proteins can then be separated from the DNA by allowing
them to settle to the bottom of a test tube (biologists are usually in
a hurry, so a **centrifuge** is used speed up the settling process, as
shown later: Figure 3-7). The DNA-containing solution can then be
removed from the tube, leaving much of the protein behind. Proteins
that remain with the DNA can be broken down by treating the
mixture with proteases (enzymes that specifically cut proteins).
Sometimes the watery mixture is shaken with phenol, an oily com-
pound that separates from water when allowed to stand. During
shaking, many contaminants move into the phenol, and they are

removed when the phenol is separated from the DNA-containing solution. Addition of alcohol to a solution of DNA causes the DNA strands to precipitate. If alcohol is gently layered on a concentrated DNA solution, the DNA precipitate forms at the boundary between the alcohol and the DNA solution. A glass rod can then be stuck into the mixture and rotated between the fingers. When the rod is lifted out of the solution, the DNA that has been spooled onto it looks much like nasal mucus. DNA that will not be used directly can be dried and stored. Dried DNA can be dissolved in a dilute salt solution, where it is then manipulated with enzymes, many of which are commercially available.

PERSPECTIVE

In 1953 James Watson and Francis Crick made what is considered to be one of the most important scientific findings of modern biology. They proposed a structure for DNA that has guided the thinking of biologists to the point that the chemistry of heredity can be clearly explained and manipulated. Determining the information content, the nucleotide sequence, of DNA molecules is routine, and the sequence for many small DNA molecules is completely known. It is taking longer to determine the sequence of the larger molecules, such as bacterial and human chromosomal DNAs, but the progress is rapid. In 1995 the complete nucleotide sequence of DNA from the bacterium called *Hemophilus* was reported (1.8 million nucleotide pairs).

Our ability to determine the nucleotide sequence of DNA from an organism emphasizes that the sequence must be quite stable. But we should not view DNA sequences as static, for they do change through errors introduced during reproduction, through the movement of small, specialized **sequences** that hop within and between DNA molecules, and through sequence exchange with other DNAs. Those changes in nucleotide sequence are the raw material on which the forces of evolution act. Indeed, every nucleotide sequence we determine today reflects the long evolutionary history of a specific DNA molecule.

DNA does not exist inside the cell as a static, naked molecule: it is constantly interacting with proteins, many of which have a profound effect on how the DNA twists and loops. Although these structural changes do not directly influence the information content—that is, the sequence of nucleotides—they do affect how other proteins interact with DNA. Thus, chromosomal proteins affect the use of genetic information by the cell, a topic discussed in Chapter 4.

Questions for Discussion

1. What is a helix and why is the DNA molecule called a double helix?
2. Genetic information is encoded in the sequence of nucleotides in DNA, and early in the next century biologists may know the nucleotide sequence for the human genome. Will this information tell us everything we need to know to cure human genetic diseases?
3. Hydrogen bonds are weak chemical interactions that contribute to complementary base pairing. If there are an equal number of A · T and G · C base pairs in a DNA molecule 4 million base pairs long, how many base-pairing hydrogen bonds would there be in the molecule? (Hint: See Figure 2-5.)
4. Atoms sometimes carry electrical charges. Two atoms of different charge (plus and minus) attract, and atoms with the same charge (both plus or both minus) repel each other. The phosphates in DNA are negatively charged. What effect does this polarity have on the ability of double-stranded DNA molecules to maintain a double-stranded structure? How might the **binding** of positively charged proteins to DNA affect the ability of DNA to exist in the double-stranded configuration?
5. The genetic code is read as triplets, and often several triplets **encode** the same amino acid (see Figure 2-6). Three triplets also signal the end of a protein (labeled TERM in Figure 2-6). Show how a single base change can convert codons for tyrosine,

cysteine, tryptophan, leucine, and lysine into termination codons.

6. Many of the activities of nucleic acids depend on their ability to recognize each other. Recognition is based on the principle of complementary base pairing (see Figure 2-5). Thus base pairing must be a property common to all living organisms. Base-pairing forces are not very strong, and temperatures near 100 °C are sufficient to break the base pairs. This causes double-stranded nucleic acids to become single-stranded. With these points in mind, explain how some organisms can grow in thermal pools at temperatures near the boiling point of water. (Hint: Question 4 may be helpful.)

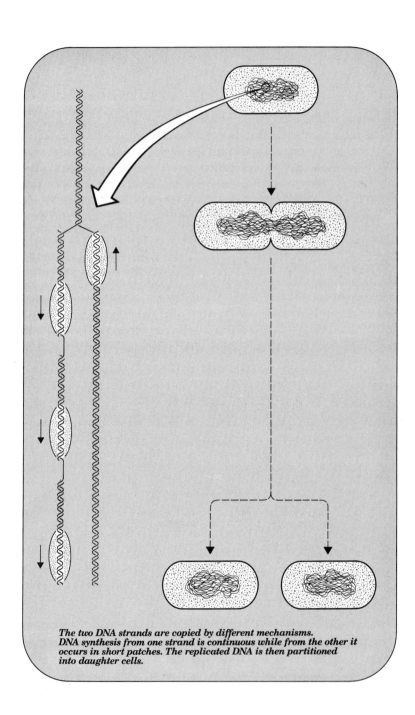

The two DNA strands are copied by different mechanisms. DNA synthesis from one strand is continuous while from the other it occurs in short patches. The replicated DNA is then partitioned into daughter cells.

REPRODUCING DNA

Information Transfer from One Generation to the Next

Overview

DNA is copied by a group of proteins traveling together along a double-stranded DNA molecule. As the proteins move along the DNA, they separate the two DNA strands and make a new strand adjacent to each old one; one double-stranded molecule becomes two. Each DNA has one old and one new strand, and the nucleotides in these two strands are again complementary. The cell controls the number of DNA molecules produced by allowing replication, as the copy process is called, to start only at specific places on DNA and only when required proteins are at the proper level. Occasionally errors occur while the new strands are being made, and the nucleotide sequence in the new strand is not perfectly complementary to that in the old strand. If not corrected, these errors, called mutations, are passed faithfully from one generation to the next.

The two strands in a DNA molecule are not replicated in exactly the same way because they run in opposite directions. Replication of one old strand occurs by formation of one long, continuous complementary strand. Replication of the other, however, occurs by formation of short patches, which are later connected by an enzyme called DNA ligase.

Enzymes involved in DNA replication have been purified, and some are important tools for gene cloning. For example, DNA ligase is used to join DNA fragments together.

INTRODUCTION

Earlier we defined life as a set of chemical reactions organized in a way that permits the set to reproduce itself. The preceding chapters touched on how DNA and enzymes organize the reactions. Now we examine reproduction itself. For the present discussion, reproduction is defined as the duplication of the information content of the cell, followed by segregation of this information into two newly formed daughter cells. Since information is stored in DNA, knowing how DNA duplicates is crucial to understanding reproduction. The next section discusses the machinery responsible for **DNA replication,** the technical term for reproduction of DNA. The following section addresses how the cell controls DNA replication to assure that the proper number of DNA copies are present. Since understanding mutations has been important for deciphering how DNA functions, the chapter continues with a section on DNA structure and mutation. A final section, outlining how enzymes are obtained and handled, prepares the way for subsequent discussions of DNA manipulation, which entail the use of enzymes involved in DNA replication.

DNA REPLICATION

To provide genetic continuity from one generation to the next, DNA not only must be chemically stable, but it also must be copied accurately during replication. If this did not happen, the DNA of an offspring would contain information different from that of its parents. Then the proteins in the offspring would differ from those in the parents, and the characteristics of the two generations would no longer be the same. Accurate copying is accomplished in the following way. The two strands of DNA separate, permitting each to act as a template for making a new strand. Nucleotides are aligned along each DNA strand according to the complementary base-pairing rule (Figure 2-5), and they are joined to form a new DNA strand. Two DNA molecules arise from one, and they contain identical information (Figure 3-1). The two DNA molecules move to different parts of the cell, cell division occurs between the DNA molecules, and two daughter cells arise having identical DNAs (Figure 3-2).

Since biochemists can replicate DNA in test tubes by adding a small number of purified components, much is known about the replication machinery. The process can be divided into a number of steps. First, specific proteins form a complex with DNA at a specific site in the DNA called an **origin of replication;** the process by which this occurs is discussed in more detail in the next section. Then the two DNA strands begin to unwind, producing a **replication**

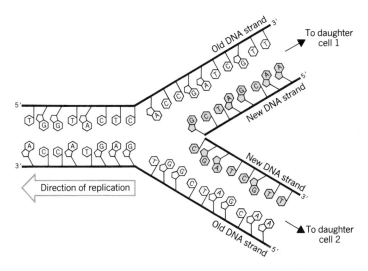

Figure 3-1 Two DNA Molecules Arise from One. Base-pairing complementarity allows information to be copied exactly. Notice that each daughter cell will receive a DNA molecule having exactly the same nucleotide sequence. For clarity, the strands are not drawn as an interwound helix (see Figure 2-1), and hydrogen bonding between complementary base pairs (Figures 2-3 and 2-5) is omitted.

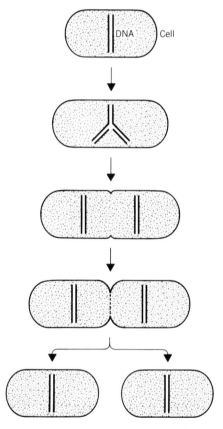

Figure 3-2 Replication and Segregation of Bacterial DNA. While the cell is growing larger, its DNA replicates. The two daughter DNA molecules move to opposite parts of the cell. A new cell wall forms between the DNA molecules, and two daughter cells are produced. A similar process occurs in the cells of higher organisms. However, they have many DNA molecules, each with many origins of replication. Consequently, the sorting system is more complex.

fork that moves through the double-stranded DNA molecule as replication occurs. The process is much like unzipping a zipper (Figure 3-1). Thus two single strands are gradually created, exposing the bases. An enzyme called **DNA polymerase** binds to one of the single strands and moves in the direction of fork movement, closely following the zipper (Figure 3-3). As it moves along the single strand, DNA polymerase mediates formation of base pairs between free

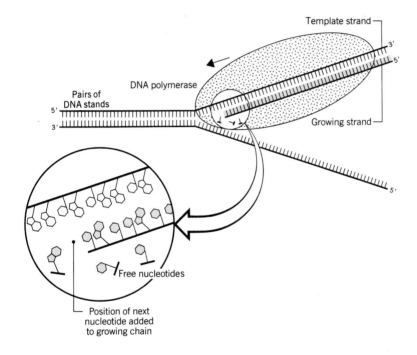

Figure 3-3 Function of DNA Polymerase. The DNA strands separate as DNA polymerase and other proteins bind to the DNA. In the replication of one strand, DNA polymerase follows the replication fork, forming a new strand whose nucleotide sequence is complementary to that of the old (template) strand. Nucleotides are added to the end of the growing chain one at a time. The order of the nucleotides in the growing chain is determined by the order in the template strand. For clarity, the DNA is drawn as two parallel strands rather than as a double helix as in Figure 2-1, and hydrogen bonding between base pairs (Figure 2-3) is omitted.

nucleotides (links not yet in a chain) and the linked DNA nucleotides. The alignment obeys the complementary base-pairing rule, so wherever an A occurs in the single DNA strand, a T is aligned opposite it. As soon as a free nucleotide has lined up at the end of the growing chain (Figure 3-3), and opposite to its complement in the template strand, DNA polymerase links it to the new chain. Then the polymerase moves down the chain one position, aligns the next nucleotide, and links it to the growing chain. Thus a new single-stranded DNA chain is formed using the information contained in the old one. As the new chain forms, it is already base-paired with the old one; a double-stranded DNA molecule containing one new strand and one old strand is produced.

A particularly interesting detail in the scheme just described is that DNA polymerase always travels in the same direction along a DNA chain (5′ to 3′ in the growing strand). Like a motion picture film, DNA has directionality, and the polymerase recognizes the directionality. This aspect of structure is important because the two strands in a double-stranded DNA molecule run in opposite directions (see Figure 2-4*b*). As the replication fork moves through the DNA molecule, unzipping the DNA and making single strands available for replication, DNA polymerase on one strand will follow the fork (Figure 3-4), continuing in the same direction as the fork. Polymerase on the other strand, however, must move in the opposite direction, away from the fork. Synthesis opposite to fork movement occurs in short patches, which in bacteria are about 1000 nucleotides long (see frontispiece, Chapter 3). Once the polymerase has made one patch, we could imagine that it leaps back toward the vee in the fork and begins making another piece of DNA until it runs into the patch of new DNA it had just laid down. Then the polymerase must again catch up with the moving fork. Since it is unlikely that DNA polymerase actually leaps toward the moving fork, molecular biologists are working on ideas involving two polymerases bound together in such a way that no leaping occurs.

An important point to emerge from the study of replication forks is that DNA polymerase is unable to join the two patches of new DNA together. That requires another enzyme, called **DNA ligase.** Genetic engineers have come to view ligase as an important tool for joining DNA fragments together.

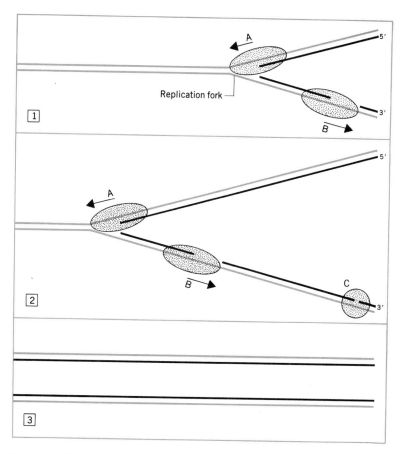

Figure 3-4 Discontinuous DNA Synthesis. (1) One DNA polymerase complex **(A)** moves continuously along an old single strand (shaded) synthesizing a new strand (solid) in the direction of replication fork movement while the other **(B)** moves in the opposite direction. **(2)** Polymerase **B** synthesizes short patches of DNA. Small gaps are filled in by another type of DNA polymerase, and the DNA fragments are joined by DNA ligase **(C)**. **(3)** The result is two double-stranded DNA molecules.

INITIATION OF DNA REPLICATION

DNA molecules are able to replicate only if they have a special site at which replication can start. These sites are the origins of replication mentioned above. They often contain short, repeated sequences that are recognized by proteins involved in the initiation process. These proteins, which are called initiator proteins, bind to the origin and distort the DNA to allow other replication proteins to bind. The complex of proteins forces the DNA strands to separate over a short region, and that modification allows still other proteins access to the DNA (some proteins bind only to single-stranded DNA). Then the key step of priming occurs. It turns out that DNA polymerase cannot begin making DNA from a template strand alone—it needs to add nucleotides to the end of a preexisting piece of nucleic acid. That preexisting piece is called a **primer.** In many cases a piece of RNA is synthesized to be a primer, since the enzymes that make RNA can do so without themselves requiring a primer. Once priming has occurred, DNA polymerase attaches and begins its job of replication.

Cells carefully control DNA replication to coordinate the number of DNA molecules with the cell cycle. This assures that when cell division occurs, one copy of DNA is available for each daughter cell. Control is exerted at the initiation step—once initiation has occurred, elongation and termination proceed without interruption. Different cell types have different strategies for initiating replication. For example, bacterial DNA molecules tend to have a single origin of replication. Two replication forks emerge from that region and travel in opposite directions around the large circular DNA to a region called the **terminus of replication,** which is roughly 180° opposite the orgin. Eukaryotic chromosomes have much more DNA to replicate during each cell cycle, and so they initiate replication from many origins scattered along the DNA. For reasons that are not completely understood, eukaryotic origins are less precisely defined than bacterial origins.

Cells are occasionally invaded by foreign DNA molecules (plasmids and viruses; Chapter 6) that use host replication machinery to replicate their DNA molecules. Each type of DNA molecule has its own specific origin and generally encodes its own initiator protein. This allows a cell to accommodate a large number of different **repli-**

cons, the technical term for DNA molecules that are capable of replication.

DNA STRUCTURE AND MUTATIONS

Occasionally errors are made during DNA replication. Those errors may be passed on to the next generation of cells, sometimes with serious consequences. Examination of the genetic code (Figure 2-6) reveals how specific nucleotide changes in DNA can lead to specific amino acid changes in protein. The end result of one such series of changes is illustrated in Figure 3-5. In an offspring a single nucleotide pair might be changed; for example, an A · T pair in the gene might be converted to a G · C, C · G, or T · A pair. In the example in Figure 3-5 an incorrect amino acid is inserted into the protein specified by the gene containing the error. An incorrect amino acid, in turn, can change the structure of the protein enough to inactivate it. Since inactive proteins frequently alter the chemistry of cells and organisms, it would not be surprising to find that the offspring looked different from the parent or acted in different ways. Such an altered organism is called a **mutant,** and if the mutation (i.e., the nucleotide change in the DNA) has produced an inactive form of a protein that is essential for life, the mutant organism will die. Occasionally mutation produces a better protein in the offspring, improving its ability to survive and reproduce in that particular environment. In such cases the mutant may eventually become the dominant type of organism in the population, and a small step in evolution will have occurred.

A number of different changes in normal DNA (Figure 3-6) have been discovered that give rise to mutations. In addition to replacement of a correct amino acid with an incorrect one (Figures 3-5 and 3-6b), changing a single nucleotide occasionally converts a normal triplet into a stop codon (see Figure 2-6)—then the operation of the protein synthesis machinery prematurely terminates (Figure 3-6c). If such a mutation occurs near the beginning of the gene, only a small fragment of the protein can be made, with obviously serious effects. In another type of mutation a nucleotide is lost during replication; that is, the replication machinery skips a letter (Figure 3-6d). In this case the information is thrown out of the correct reading

↗ detergent

SDS - removes lipids
from the cell membrane
encourage cell to lyse

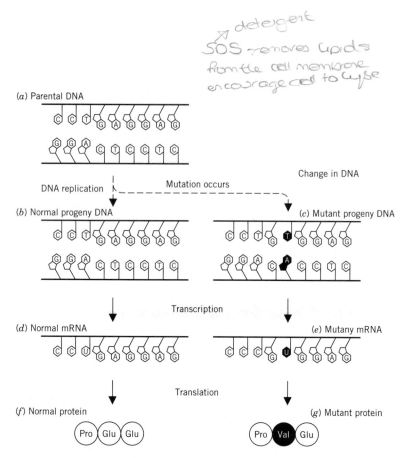

Figure 3-5 A Point Mutation Changes the Sequence of Amino Acids in a Protein. DNA replication is very accurate, so the nucleotide sequence in the progeny DNA **(b)** is identical to that of normal parental DNA **(a).** Occasionally an error is made. In this example, a particular A · T base pair in parental DNA changes to a T · A pair in the mutant, progeny DNA **(c).** During transcription, the information in DNA is converted into messenger RNA. The mutation in DNA results in a conversion of particular GAG codon in normal messenger RNA **(d)** into a GUG codon in mutant messenger RNA **(e).** During translation of the information into protein, GAG codes for the amino acid glutamic acid (Glu) **(f),** while GUG codes for valine (Val) **(g)** (see Figure 2-6). The two amino acids have very different chemical properties. Since the structure of the resulting protein is determined by the precise order of the amino acids, the mutant protein will differ significantly from the normal protein. The differences between the normal and mutant molecules shown are identical to those found between healthy people and patients suffering from sickle-cell disease.

48

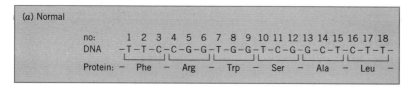

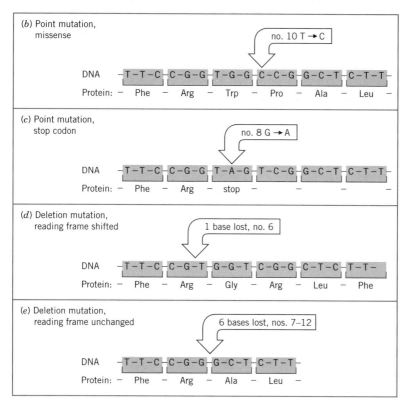

Figure 3-6 Common Types of Mutation. (a) A normal nucleotide sequence for one strand of DNA codes for a protein having the six amino acids listed (Phe, Arg, Trp, etc). The codons for the amino acids are bracketed above the respective amino acids. **(b)** If a T is changed to a C (arrow), the resulting mutant protein has a proline where serine is normally located. This type of change is called a missense mutation. **(c)** If a G is changed to an A (arrow), a stop codon is created and **protein synthesis** halts. This change is called a nonsense mutation, and the mutant protein is shorter than the normal protein. **(d)** Deletion of one base throws the reading frame out of register (frameshift mutation), and incorrect amino acids (Gly, Arg, Leu) occur in the mutant protein. **(e)** Removal of six bases produces a deletion mutation and a protein missing two internal amino acids (Trp, Ser).

49

frame, and many incorrect amino acids are placed in the protein. This is one type of **frameshift mutation;** the insertion of an extra nucleotide is another. In still other cases, a large stretch of DNA is lost (Figure 3-6*e*). Fortunately, mutations rarely occur in the absence of chemical agents or radiation, and as a result organisms have stable characteristics.

The accuracy of the replication machinery is very impressive. In bacteria, detectable mutations (errors) in a given gene arise at a frequency of only one in a million cells per generation, even though the replication apparatus copies the 4 million base pairs of DNA at a rate of 50,000 base pairs per minute. The key to understanding this process lies in the principle of base-pairing complementarity. The important point is that biological molecules recognize each other by fitting together like locks and keys. Nearly perfect fits are required for two molecules to bind together, and when the fit is good, the binding can be strong. As long as the rule illustrated in Figure 2-5 is obeyed (namely, that A bind only to T and G only to C), there will be no errors and replication will proceed properly.

Within this framework, any chemical that converts one letter to another in the old DNA strand is capable of "tricking" the replication machinery into inserting the wrong nucleotide into the new DNA strand as it is made. Chemicals that alter the information in DNA are called **mutagens.** Chemical mutagens, which are everywhere in our environment, and physical mutagens such as **ultraviolet light,** have become major threats to human health. There is little doubt that both kinds of mutagen can lead to certain types of cancer. Bacteria are now being used to test household items, food additives, and pesticides for their ability to cause mutations and presumably cancer. The test is based on our ability to easily detect the creation of bacterial mutants by observing whether they grow on **agar plates** containing particular nutrients (bacterial growth is described in more detail in Chapter 5).

Mutations are also central to understanding genetic diseases, since in one sense these maladies can be attributed to mutations that are passed from one generation to the next. By knowing the particular nucleotide changes responsible for a disease-causing mutation, it is possible to identify persons prone to certain diseases. Huntington disease, for example, can be predicted decades before onset of symptoms on the basis of the result of testing a person's DNA. Human mutations are considered in more detail in Chapter 13.

HANDLING ENZYMES

Thus far enzymes have been considered to be tiny agents that somehow join nucleotides to form DNA or RNA chains, join amino acids to form protein chains, and in general direct the chemical reactions of the cell. To provide a better understanding of enzymes, the processes they control, and how they are used in genetic engineering, it is necessary to digress from describing biological principles into a discussion of some of the technical aspects of handling enzymes. Methods for obtaining DNA polymerase are outlined below to illustrate how enzymes are detected and purified.

Before enzymes can be used for gene cloning, they must be removed from living cells. The first step of **purification** is to obtain a large batch of cells, about a pound of cells in the case of bacteria. The enzymes are liberated by breaking down the cell walls: one method is to combine the cells with tiny glass beads and grind the mixture in a blender. Then a series of physical manipulations is begun to separate the enzyme of interest out of the **cell extract**—the mixture of cellular components. Separation is possible because every type of enzyme is physically and chemically different from every other type of molecule in the cell. For example, DNA polymerase is relatively small compared to many other cell components. Thus one purification step might be to allow the large cellular debris to settle to the bottom of a test tube. The settling process is greatly accelerated by centrifugation (Figure 3-7). DNA polymerase is expected to remain in the fluid at the top of the tube. A crucial part of this purification step is determining whether DNA polymerase has settled to the bottom of the tube or stayed in the fluid. Thus it is necessary to develop an **assay** for the enzyme—that is, a way to detect and follow the enzyme through a series of purification steps.

To develop an assay for a particular enzyme, a biochemist must understand the nature of the chemical reaction accelerated by the enzyme. From the discussion in the first part of this chapter, one could imagine that DNA polymerase, using a single-stranded DNA template, links free nucleotides together to form a DNA chain. Figure 3-8 shows how this concept can be used to determine whether an extract of broken cells contains DNA polymerase. Generally the number of free nucleotides that are linked together

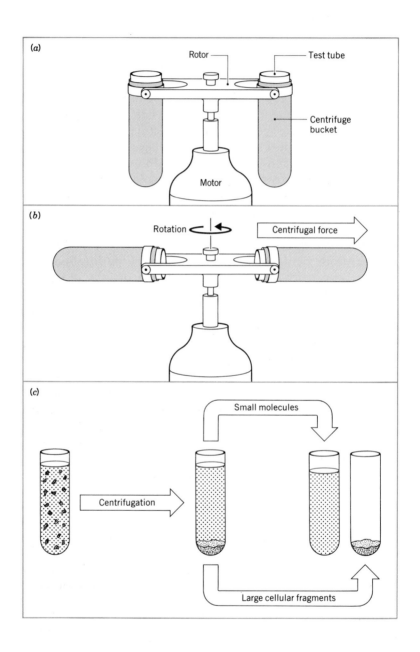

52

in a test tube reaction is very small; therefore, it is necessary to have a very sensitive measure for this conversion. Radioactive isotopes are used because very small amounts of them can be detected. (Radioactive isotopes are unstable forms of atoms that spontaneously disintegrate. Those used in biochemical studies emit high energy particles that can be detected on film or with instruments such as Geiger counters. A wide variety of radioactive molecules are commercially available.)

Operationally, a mixture is prepared consisting of a cell extract that contains the DNA polymerase, the four nucleotides, one of which is radioactive, and a single-stranded DNA template to which is attached a short complementary **oligonucleotide** primer (see Figure 3-8a; as pointed out above, DNA polymerase cannot begin DNA synthesis with a template strand alone—it must always add nucleotides to the end of a preexisting strand). After the mixture has been incubated for an hour or so at 37 °C (body temperature), cold acid is added to the mixture. DNA molecules clump together to form a white, stringy precipitate, a solid that settles to the bottom of the test tube. If the acid-containing mixture is then poured into a funnel lined with filter paper, the solution will pass through the filter paper. However, precipitated DNA sticks to the paper. Free nucleotides are relatively small, and they do not precipitate when acid is added to the mixture; consequently, they are not trapped by the filter paper. The only radioactive molecules that can stick to the filter paper are the nucleotides that have become a part of the DNA. Thus, the amount of radioactivity on the filter paper is a measure of the number

---

Figure 3-7 Fractionation of a Cell Extract by Centrifugation. (a) Schematic diagram of a swinging bucket rotor. Test tubes filled with the cell extract are placed in the centrifuge buckets. **(b)** The motor causes the rotor to spin, the buckets swing into the horizontal position, and the molecules in the cell extract sediment (migrate) toward the bottom of the test tube. Ultracentrifuges can generate forces in excess of 500,000 times gravity. **(c)** The force generated by the **centrifuge** causes molecules to separate on the basis of size and shape. After the large molecules have sedimented to the bottom of the tube, the upper solution can be carefully removed with a **pipette** and placed in a second tube.

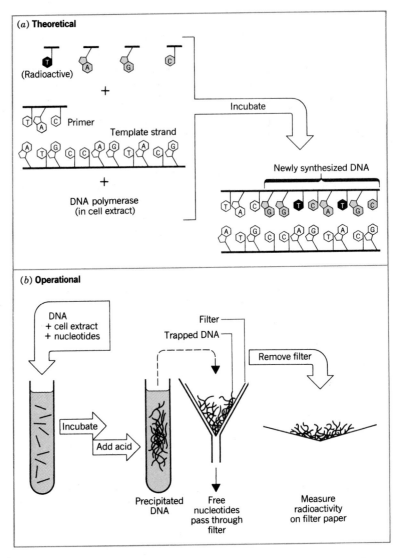

Figure 3-8 Assay for DNA Polymerase. (a) Theoretical. A mixture is prepared that contains the four nucleotides [A, T (radioactive), C, and G], a single-stranded DNA template containing a short, double-stranded region that provides a primer (DNA polymerase always requires a primer to begin its action), and a sample containing DNA polymerase. For polymerization,

of nucleotides that have been linked to form DNA, which in turn is a measure of the DNA polymerase activity present.

Biologists use many different enzymes. For each one, a different assay must be devised to follow the enzyme during purification. However, the principle is the same for each assay: there must be a way to distinguish between the reaction **substrate** (the molecules you start with) and the reaction **product** (the molecules you end up with). The biologist measures the speed at which substrate is converted into product. In the example above, free nucleotides are the substrate of DNA polymerase, and nucleotides incorporated into DNA are the product of the reaction. The speed of the reaction is determined by measuring the amount of radioactivity sticking to the filter paper per minute of incubation time at 37 °C for the total reaction mixture. The speed of the reaction indicates the amount of enzyme present in the cell extract (enzymes accelerate chemical reactions). Thus, after separating cellular components into different test tubes (Figure 3-7), the biologist uses the assay (Figure 3-8) to determine which tube contains the enzyme.

A number of methods are used to separate subcellular components. In the procedure called **column chromatography,** a glass tube is mounted vertically and filled with a solid material such as chemically modified **cellulose** powder. The cell extract is allowed to flow

---

each nucleotide must be attached to three phosphate (PO_4) groups. During incubation, DNA polymerase joins the free nucleotides together to form DNA, releasing two phosphates from each **nucleoside triphosphate.** The newly made DNA will be radioactive due to incorporation of radioactive T. (b) Operational. The mixture is added to a test tube and incubated to allow the reaction to occur. Acid is added to stop the reaction and to cause the DNA molecules to clump together. The mixture is poured through a piece of filter paper, and the clumped DNA is trapped on the paper. Any radioactivity incorporated into DNA during the reaction will stick to the filter paper because of the large size of DNA. In contrast, molecules of T (radioactive) that were not incorporated into DNA are washed through the filter because they are small. The radioactivity on the filter paper is measured with an instrument called a liquid scintillation counter.

slowly through the packing in the glass tube, eventually dripping out the bottom of the tube. Some molecules bind tightly to the cellulose (or other packing), others loosely. Thus, when the biologist passes a dilute salt solution through the packing in the tube, some molecules come out after very little salt solution has passed through, whereas others require extensive washing. By collecting the salt solution in a series of test tubes (Figure 3-9), the biologist can separate various molecules. Then the contents of each tube can be tested to determine which contains the enzyme being sought. By combining several chromatographic procedures, it is possible to separate an enzyme from all other cellular components.

PERSPECTIVE

By the early 1950s the importance of DNA to heredity was widely recognized, and in 1953 when Watson and Crick published their two-stranded helical model of the DNA molecule, biochemists had already set out to determine how DNA replication works. At the time biochemists were so successful at finding ways to purify and study new enzymes that the era has often been called the Age of the Enzyme Hunters. By the late 1950s Arthur Kornberg had found that DNA polymerase required a template DNA to make new DNA, and that discovery provided strong experimental support for the model of Watson and Crick. By the early 1970s a number of proteins that participate in DNA replication had been purified. Consequently, when methods were found to cut DNA into specific pieces, replication proteins that would join the pieces were already available. In addition, other replication proteins were known that could synthesize highly radioactive DNA to be used as probes for locating bacterial colonies containing cloned genes. Thus information from a number of lines of research was used to develop gene cloning strategies.

DNA replication occasionally introduces errors into DNA, as do environmental factors. To provide genetic stability, organisms have evolved DNA repair proteins that remove errors. The importance of mutation repair was recently emphasized by the finding

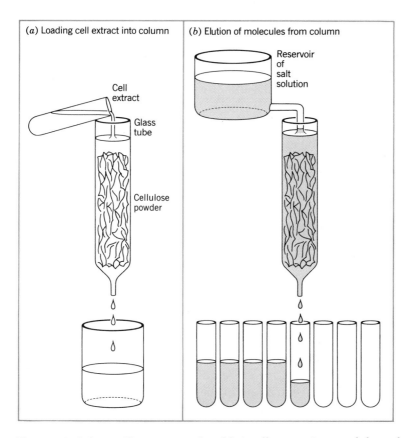

Figure 3-9 Column Chromatography. (a) A cell extract is passed through a solid matrix (e.g., chemically treated cellulose powder) to which protein molecules stick with varying degrees of tightness. **(b)** The proteins can be removed from the column by passing a dilute salt solution through the column at gradually increasing concentrations. Some molecules will come through sooner than others. All of the salt solution is collected in a series of test tubes, and the contents of the tubes are assayed to determine which tube contains the protein being sought.

that heritable defects in repair proteins can predispose a person to cancer.

Questions for Discussion

1. Why is complementary base pairing important for understanding how DNA is duplicated?
2. During the normal cycle of DNA replication, each of the two parental DNA strands is replicated; the result is two new double-stranded DNA molecules that are identical. But some viruses contain only a single strand of DNA. Others contain only a single strand of RNA. How might these nucleic acids be replicated?
3. The circular chromosome of the bacterium called *E. coli* is replicated in about 40 minutes by two forks proceeding in opposite directions from a single origin of replication to a termination point 180° from the origin. The chromosome is about 4.4 million base pairs long. If there are 10.4 base pairs per turn, calculate how fast the DNA must rotate along its long axis, in revolutions per minute, to unwind as the forks move. How might the torsional tension ahead of the forks be relieved?
4. If it takes *E. coli* 40 minutes to replicate its chromosome, how can this bacterium divide every 20 minutes, with each daughter cell getting a complete copy of the chromosome?
5. Since DNA polymerase requires a primer to begin synthesis and since one of the DNA strands is replicated as short pieces (about 1000 nucleotides long in bacteria), many priming events must occur during the synthesis of the entire chromosome. Calculate that number for a bacterial chromosome having 4 million base pairs.
6. Many chemicals in our environment cause mutations in bacteria. Describe a way in which mutation frequency might be used to test new chemicals before they reach the environment (see Chapter 5 for discussion of bacterial growth).

7. Design an experiment with living bacteria that would tell you whether a newly discovered antibiotic rapidly blocks DNA replication. (Bacteria are single-celled organisms that can be grown in liquid medium in flasks; thus uniform samples can be easily withdrawn for analysis in test tubes).

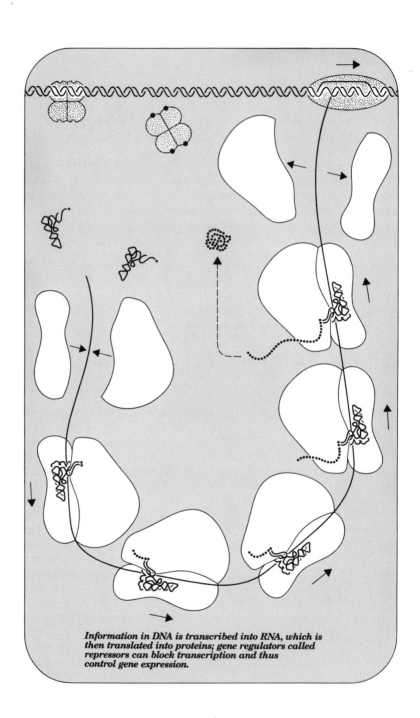

Information in DNA is transcribed into RNA, which is then translated into proteins; gene regulators called repressors can block transcription and thus control gene expression.

GENE EXPRESSION
Cellular Use of Genetic Information

Overview _____

A gene is a specific region of DNA, a specific stretch of nucleotides, that contains information for making a particular protein. Cells make thousands of different proteins, and each cell contains thousands of different genes. Gene expression is the process by which the information in DNA is converted to protein.

To make a certain protein, the cell first uses the information in a particular gene to direct the formation of a single-stranded molecule called RNA. This process is called transcription. RNA and DNA are structurally similar, and both use a 4-letter alphabet to store information. The nucleotide sequence of any given RNA molecule is determined by the nucleotide sequence of the gene used to make the RNA. One type of RNA is called messenger RNA, and the information it contains is used to direct the formation of protein by a process called translation. In this process messenger RNA first binds to subcellular workbenches called ribosomes. It then feeds across the ribosomes, and as it does, amino acids, the subunits that make up proteins, align in the order specified by the nucleotides in the messenger RNA. During the alignment process, amino acids are linked together to form a protein chain.

The alignment of amino acids involves a second type of RNA molecule called transfer RNA. Transfer RNA molecules are also encoded by genes and are made in the same way as messenger RNA. Transfer RNA molecules serve as adapters, converting the 4-letter alphabet of DNA and RNA into the 20-letter alphabet of proteins. Cells contain at least one transfer RNA type for each of the 20 amino acid species that make up proteins. One end of a transfer RNA attaches to a specific type of amino acid, while another part attaches to a specific codon in the messenger RNA. The transfer RNAs line up along the messenger RNA in the precise order dictated by the information from the gene. Since each transfer RNA is also attached to one of the amino acids, ordering the transfer RNAs also orders a series of amino acids.

Mechanisms have evolved to control the timing of gene expression, that is, to dictate when a given protein is made from the information stored in its gene. In some cases specific proteins bind to a gene and block its expression. In other cases proteins enhance expression. Since each regulatory protein has its own gene, one gene can influence the expression of another. Sometimes a protein product of a gene influences expression of many other genes, including its own. Elaborate circuits of gene control exist, and they allow a cell to respond quickly to changes in its environment. Successful genetic engineering often requires that these regulatory mechanisms be understood.

INTRODUCTION

Gene expression is the process whereby information stored in DNA directs the construction of RNA and proteins. In this process information is first transcribed into RNA, a single-stranded, DNA-like molecule. The 4-letter nucleotide code is then translated from RNA to the 20-letter amino acid language of proteins. Gene expression is tightly controlled, and it plays a major role in determining the amount of each type of protein a cell contains. The selective control of gene expression allows multicellular organisms to have different cell and tissue types, even though almost every cell in an organism contains the same genetic information.

The expression of a gene involves many biochemical reactions; consequently, control mechanisms can act during many steps in the process. Specific proteins bind to DNA and selectively raise or lower synthesis of messenger RNA. Other factors, some of them proteins, affect how quickly messenger RNA is broken down into nucleotides. Even binding of ribosomes to the messenger is subject to control. Determining how all the regulatory factors contribute to the tissue-specific control of human genes is one of the major challenges facing molecular biologists. Several gene control mechanisms are outlined near the end of the chapter to solidify the principles that have been introduced and to provide an appreciation for the intricacies of molecular interactions.

TRANSCRIPTION

Transcription is the synthesis of an RNA molecule using a DNA molecule as a template; genetic information is converted from a DNA form into an RNA form. RNA is a long, chainlike molecule similar to DNA, but it differs slightly in the sugar part of the back-bone: RNA uses the sugar ribose, which is similar to deoxyribose (see Figure 1-4), but ribose has an oxygen attached to the 2' carbon as well as to the 3' and 5' carbons. In addition, RNA is shorter than DNA, and as pointed out earlier, it contains the base uracil (U) instead of thymine (T). In the process of transcription, an enzyme, **RNA polymerase,** recognizes and binds to a DNA nucleotide se-quence just before the beginning of the gene. This **recognition site,** called a **promoter,** positions RNA polymerase properly on the correct DNA strand and points it in the right direction (the strands of DNA are complementary, not identical, and they run in opposite directions). The RNA polymerase–DNA interaction causes the DNA strands to separate over a short distance, and the polymerase moves into the gene (Figure 4-1). As RNA polymerase moves, it creates a new chain by linking together individual ribonucleotides existing free in the cell. The order of the nucleotides in the new RNA chain is determined by the complementary base-pairing rule. If the first letter RNA polymerase encounters in DNA is a T, the enzyme will add an A to the chain it is making. Likewise, if the next DNA letter is G, a C will be added to the new chain. Eventually a stop signal at the end of the gene is reached, RNA polymerase comes off the DNA, and the new RNA chain is released. The new chain is called messenger RNA because it carries information from DNA to sites

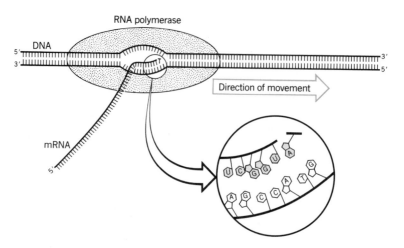

Figure 4-1 Transcription. The enzyme complex called RNA polymerase causes the DNA strands to separate over a short region (10–20 base pairs). The polymerase moves along the DNA, and as it does, it forms an RNA chain using free nucleotides. The order of the nucleotides in RNA is determined by the order of nucleotides in one of the DNA strands through the complementary base-pairing rule. In this example, the nucleotide sequence of the RNA is complementary to that of the lower DNA strand. For simplicity, the DNA strands are not drawn as an interwound helix (see Figure 2-1), and complementary base pairing is not shown.

where the information is used to specify the order of amino acids in proteins. Other RNA molecules, such as transfer RNA and ribosomal RNA, are also encoded by genes, and they are made by the same process as messenger RNA.

TRANSLATION

Messenger RNA transports information from DNA to subcellular structures called **ribosomes.** Ribosomes are large (in molecular terms) ball-like structures composed of special RNA molecules (ribosomal RNA or rRNA) and ribosomal proteins. It is on the ribosomes that information in messenger RNA undergoes **translation** from the nucleotide language into the amino acid language. As a chain of amino acids is made, it spontaneously folds to form the protein specified by the gene in the DNA.

The translation machinery works in the following way. A ribosome attaches to the messenger RNA near a site on the messenger called the start codon, a three-base triplet that indicates where to start reading the message (Figure 4-2). In bacteria, translation begins before messenger RNA synthesis has been completed; thus, both messenger RNA and ribosomes can be attached to DNA simultane-

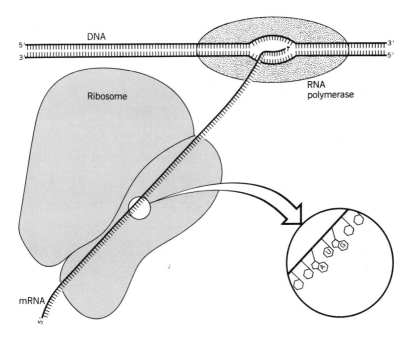

Figure 4-2 Schematic Representation of Messenger RNA Being Formed by RNA Polymerase and Attaching to a Ribosome. The ribosome, composed of two large RNA–protein subunits, binds to messenger RNA. Ribosome–messenger binding requires that a particular transfer RNA also bind to the AUG (or in some instances GUG) codon on the mRNA. This transfer RNA (not shown) is attached to the amino acid destined to become the first in the new protein chain (see below: Figure 4-4). In bacteria, messenger RNA is still attached to DNA when it binds to as ribosome, as shown. In more advanced organisms, such as humans, the messenger RNA is released from the DNA and travels out of the nucleus before attaching to a ribosome.

ously (Figure 4-2). In a human cell, DNA is located in the **nucleus,** while ribosomes are in the **cytoplasm.** Consequently the messenger must leave the DNA before attaching to ribosomes. Amino acids, which eventually will be linked to form a protein, are normally free in the cell. They are brought to the ribosomes joined to **transfer RNA** (tRNA). To accomplish the joining, each type of amino acid is recognized by a special type of enzyme called an **aminoacyl–tRNA synthetase** (Figure 4-3*a*). There are more than 20 different types of

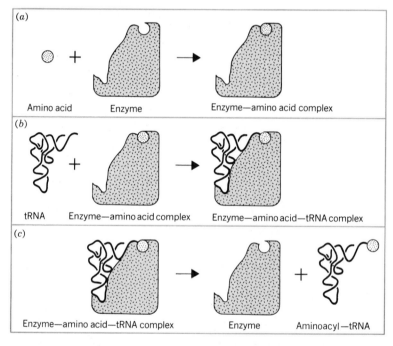

Figure 4-3 Symbolic Representation of an Amino Acid Joining to a Transfer RNA. (a) An aminoacyl–tRNA synthetase (labeled Enzyme) recognizes and attaches to an amino acid. Each of the 20 amino acid types is recognized by a different aminoacyl–tRNA synthetase. **(b)** The enzyme then recognizes and binds to a specific type of transfer RNA (there are more than 20 different types of transfer RNA molecule, at least one for each type of amino acid). In the process, the transfer RNA and the amino acid are joined to form an aminoacyl–tRNA. **(c)** The aminoacyl–tRNA is released from the enzyme, which is then free to repeat the process.

aminoacyl–tRNA synthetase, at least one for each type of amino acid. Each of these enzymes recognizes and attaches to only one type of amino acid. Each enzyme is also able to recognize and attach to a specific type of transfer RNA. There is a different type of transfer RNA for each type of amino acid. Once a particular amino acid and a particular transfer RNA have attached to a particular aminoacyl–tRNA synthetase, the synthetase links the amino acid to the transfer RNA (Figure 4-3*b*). The amino acid–transfer RNA pair is then released from the enzyme (Figure 4-3*c*). The net effect is to create an amino acid–transfer RNA pair for each amino acid type.

Each of the transfer RNAs has a sequence of three nucleotides called an **anticodon** (CAU in Figure 4-4) that is different in each type of transfer RNA. Thus, the particular amino acid at one end

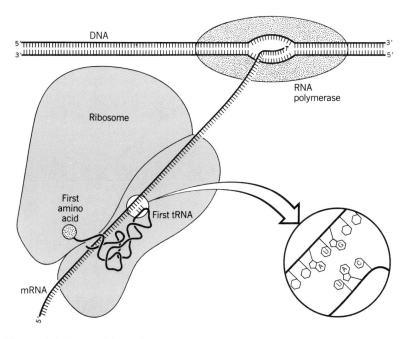

Figure 4-4 Recognition of Codons by Anticodons. The start codon (AUG) on the messenger RNA and the anticodon (CAU) of the first transfer RNA bind on the ribosome. The amino acid destined to be first in the new protein chain is already attached to the first transfer RNA.

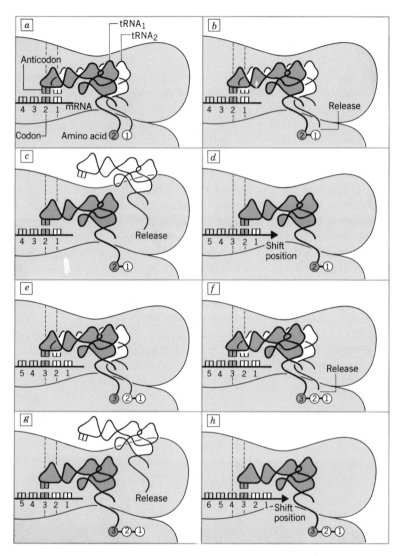

Figure 4-5 Ordering Amino Acids During Protein Synthesis. (a) After the messenger RNA (mRNA), the first aminoacyl–tRNA, and ribosome have formed a complex (Figure 4-4), a second tRNA, with its attached amino acid, is ordered on the ribosome when its anticodon region base-pairs with the second codon of the mRNA. **(b)** Amino acids 1 and 2 are joined; amino acid 1 is released from tRNA 1 (note break). **(c)** tRNA 1 is released from the ribosome. **(d)** mRNA and tRNA 2, now attached to two amino acids, are translocated (shifted over one position on the ribosome). This brings codon 3 into position on the ribosome. **(e)** Aminoacyl–tRNA 3 attaches to

of the transfer RNA always corresponds to a specific set of three nucleotides in the anticodon region of the transfer RNA. The specificity of the aminoacyl–tRNA synthetase ensures that this is the case. The anticodon on the transfer RNA can be exposed to form base pairs with the messenger RNA (Figure 4-4). One particular transfer RNA has an anticodon triplet complementary to the start codon, or triplet, on the messenger RNA. That transfer RNA and the messenger RNA lock together on the ribosome so that the two triplets, the codon on the messenger and the anticodon on the transfer RNA, form base pairs. This joining is governed by the complementary base pairing rule: if the start codon on the messenger RNA is 5′ AUG 3′, the only transfer RNA that will fit has an anticodon that reads 5′ CAU 3′ (see Figure 4-4; the 5′ and 3′ designations indicate directionality in the two strands, as shown in Figures 1-4*d* and 2-4). The particular amino acid attached to this transfer RNA is destined to become the first link in the new protein chain.

Figure 4-5 illustrates how the amino acids are ordered in the new protein. The second codon on the messenger is also locked into place on the ribosome next to the first codon. It, too, is recognized by the anticodon of a transfer RNA molecule carrying an amino acid, the amino acid destined to become the second link in the new protein. The ribosomal RNA and proteins attached to the ribosome then help join the two amino acids together. The first amino acid separates from its transfer RNA, and that transfer RNA separates from the messenger, completing one cycle of the translation process. The messenger now feeds across the ribosome much as a magnetic tape runs over the player head of a tape recorder. One after another the codons are locked into place on the ribosome. The appropriate transfer RNA binds to each triplet, placing the correct amino acid in position to be joined to the growing protein chain. When the stop signal comes along, the

---

the ribosome and forms base pairs with codon 3. (**f**) Amino acid 3 is joined to amino acid 2, repeating step b. (**g**) tRNA 2 is released from the ribosome, repeating step c. (**h**) tRNA 3 and the growing protein chain are translocated, repeating step d. This process continues until a stop codon is reached. At this point the protein chain is released from the last tRNA.

messenger falls off the ribosome. The new protein is released into the cell, and it begins to control the specific chemical reaction for which it was designed.

In summary, information from the DNA, encoded by four different letters, is first transcribed into a message (messenger RNA), using a similar 4-letter code. The message then binds to a ribosome. Small RNA molecules, called transfer RNAs, serve as adapters to convert the 4-letter alphabet of DNA and RNA into the 20-letter alphabet of proteins. During protein synthesis, the transfer RNA molecules move amino acids into position along messenger RNA at a point where the message is feeding across the ribosome. There the amino acids are linked together to form a protein chain. All organisms on our planet use the same process for making proteins, leading biologists to conclude that life is a continuum.

POSTTRANSLATIONAL MODIFICATION

Frequently proteins undergo changes after being synthesized. In some cases the protein does not become active until a portion is cut off. By using this strategy the cell can accumulate the inactive form, and when large amounts of active protein are needed quickly, a simple clipping or phosphate addition will bring about the desired activity. In other cases, the protein needs to move to a specific compartment in the cell. The various compartments are generally bounded by membranes, and so a protein must pass through a membrane to reach its destination. Regions called **signal peptides,** which are located on the amino-terminal end of the protein, help certain proteins move across membranes. As this movement occurs, the signal peptide is cut off.

Multicellular organisms also need to move proteins from one location in the body to another, and for this address labels are placed on the protein. These labels sometimes take the form of sugar molecules that are added to the protein after it is made. The process, which is called **glycosylation,** is sometimes needed for the proteins to be active and to protect the protein from degradation.

CONTROL OF GENE EXPRESSION: REPRESSION

Organisms have intricate mechanisms for controlling when genes are turned on and turned off, that is, when the information in a specific gene will be used to synthesize mRNA and when this process is prevented. These control mechanisms allow microorganisms to adapt very quickly to changes in the environment, and they allow the cells of higher organisms to develop into a variety of complex structures. Molecular biologists are trying to determine in great detail how cells manage to regulate gene expression, since this understanding is greatly improving our ability to insert genes into bacteria to produce large quantities of a specific protein, into humans to correct defective genes, and into plants and animals to improve food sources.

Biologists have found that nature has a number of ways to regulate gene expression. In the type called **repression,** RNA synthesis is blocked by a particular protein called a **repressor.** The repressor binds specifically to DNA just in front of the gene it controls (Figure 4-6),

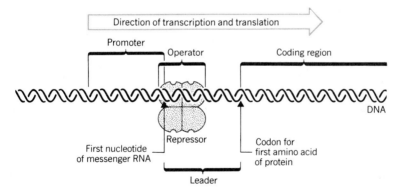

Figure 4-6 Control of Gene Expression by a Repressor. RNA polymerase normally binds to a region of DNA called a promoter. The polymerase then makes a short leader RNA followed by the coding region of the gene. The stretch of DNA that has the information for the RNA leader also serves as a binding site for the repressor. This stretch of DNA is called the operator. When the repressor binds to the operator, it blocks RNA polymerase from binding to this region of DNA and thus prevents synthesis of the messenger RNA.

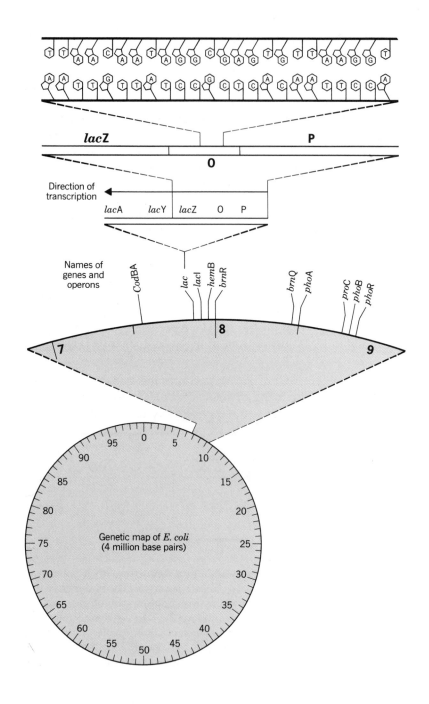

at a spot called the operator. As long as the repressor sits on the DNA, RNA polymerase is unable to start producing a message from that gene. Gene control by repressors has been most thoroughly studied with genes that code for enzymes involved in the breakdown of sugars entering bacterial cells. Unless a particular sugar is present, there is no point in producing the enzyme that breaks it down. If the sugar suddenly becomes available and enters the cell, the repressor binds to the sugar; then, since the repressor–sugar complex is unable to bind to DNA, repression is lost. As soon as the mechanism blocking RNA synthesis (i.e., repression) ceases to operate, RNA polymerase binds to the promoter, which in bacteria is generally located near the beginning of the gene. The polymerase then promptly makes the messenger for the enzyme required to break down the sugar. As the messenger is being made, it attaches to ribosomes and is translated, as described earlier in this chapter, producing the degradative enzyme. And as soon as the enzyme has been made, it begins to break down the sugar until none is left for binding to the repressor. The repressor then reattaches to its spot on the DNA, halting production of the messenger for the degradative enzyme. Thus, the cell produces the specialized degradative enzyme only when that particular enzyme is necessary.

In bacteria, several genes in a row are sometimes transcribed as a part of a single, long RNA molecule. Such a unit of genes is called an

---

Figure 4-7 Organization of an Operon in *E. coli.* The *lacZ, lacY,* and *lacA* genes are involved in the metabolism of lactose (milk sugar). Together the three genes make up the *lac* operon, and they are transcribed into a single messenger RNA molecule. The promoter **(P)** is the region where RNA polymerase binds, the operator **(O)** is the region where the repressor binds, and *lacZ* is the gene for the enzyme that degrades lactose. The exact sequence of DNA nucleotides is known for the lactose gene region; the nucleotide sequence for the binding of the lactose repressor protein to the DNA and the relationship of this stretch of DNA to the chromosome are shown. An *E. coli* DNA molecule is a large circle that, if stretched out, would be 1000 times longer than an *E. coli* cell. The genes in this DNA molecule are arranged in a circular map divided into 100 units. An enlargement of the region of the map near position 8 shows the location of the *lac* operon. For clarity, the DNA strands are not shown as a double helix, and hydrogen bonds between bases are omitted.

operon. By controlling transcription at the beginning of the operon, the repressor regulates production of several genes at once. One of the best-studied operons is called *lac,* whose three genes are involved in the transport and breakdown of lactose (milk sugar). Figure 4-7 shows how these genes, *lacZ, lacY,* and *lacA,* are arranged on the bacterial chromosome (gene names are usually italicized). RNA polymerase begins transcribing RNA from the site labeled P in Figure 4-7 and stops after passing through the *lacA* gene. Thus, the information from three genes is transcribed into a single RNA molecule. When the *lac* repressor (encoded by the nearby *lacI* gene, Figure 4-7) binds to the *lac* **operator** (region O in Figure 4-7), transcription is simultaneously blocked for all three genes. Part of the operator nucleotide sequence is shown in Figure 4-7.

CONTROL OF GENE
EXPRESSION: ATTENUATION

Attenuation is a mechanism for controlling gene expression in which the synthesis of messenger RNA is halted after only a short portion of it has been made. This process has been most extensively studied with bacterial genes involved in making amino acids. Amino acids are essential for the health of the cell as building blocks for proteins, so they need to be kept in constant supply. However, their production costs the cell a considerable amount of energy. As a result, mechanisms have evolved that carefully control amino acid production and maintain the correct balance of the 20 different amino acids. Attenuation is one of these mechanisms.

In the attenuation of the **tryptophan** genes, RNA polymerase begins making RNA some distance from the beginning of the first gene in the operon, thus creating a **leader** RNA (Figure 4-8). The leader RNA contains a coding region for a short leader protein, and some of the codons in this region specify that tryptophan is to be inserted into the leader protein. Also within the RNA leader is a variably active stop signal called an attenuator. When tryptophan is abundant, ribosomes are able to translate the leader protein. When this happens, the attenuator region of the RNA stops RNA polymerase movement by folding into a structure in which two regions of the RNA are held together by complementary base pairing. Consequently, only the short leader message is made. When tryptophan is scarce, the ribosomes stall when they come to the tryptophan codons in the message for the leader protein. The stalled ribosomes sit on the RNA and cause the RNA to fold in a different way, one that allows RNA polymerase to continue down the DNA. Then the entire message is made for the proteins involved in tryptophan production.

CONTROL OF GENE EXPRESSION: ACTIVATION

A general strategy for turning on genes involves the binding of special proteins near promoters to facilitate the proper binding of RNA polymerase. Gene activation appears to be particularly widespread among genes in complex organisms, where large numbers of related genes must be regulated together to ensure the development of specific cell types at the correct times.

In higher cells, cases of gene activation involving several different proteins and several regions of DNA have been found. Usually these control regions lie upstream from the start of the coding region of the gene, and often multiple regions are sites for the binding of specific proteins (transcription factors). Some of these proteins are general activators; that is, they are found in many cell types. Others are found only in certain cell types, such as liver or kidney. When the correct constellation of control proteins is present, the gene becomes activated and its protein product is made. Of course the control proteins are themselves produced by genes that are subject to control, so the regulatory networks can be quite complex. In certain cases a regulatory protein activates its own gene as well as other genes. Thus once the regulatory protein has been made in a particular type of tissue, its self-activation assures permanent activation for the life of the organism. Such a phenomenon could contribute to the differentiation of our cells into specific types.

DNA regions important for the control of a gene can be located far from the gene. Small segments called **enhancers** can activate genes even when the enhancer is a thousand nucleotides from the gene. Some enhancers are located upstream or downstream from a gene or in the middle of it. Enhancers appear to be binding sites for specific proteins that in turn bind to other proteins attached to re-

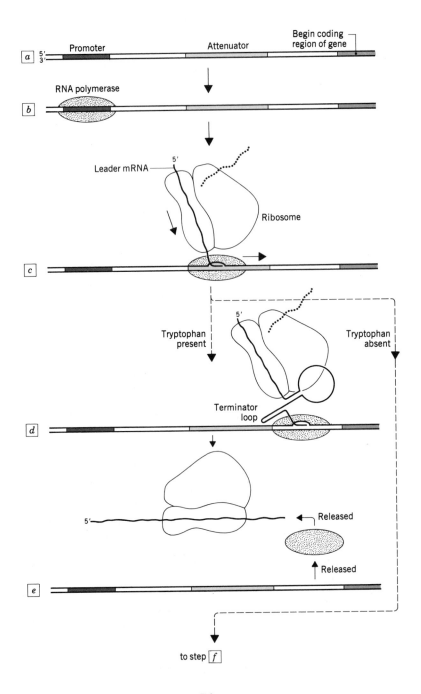

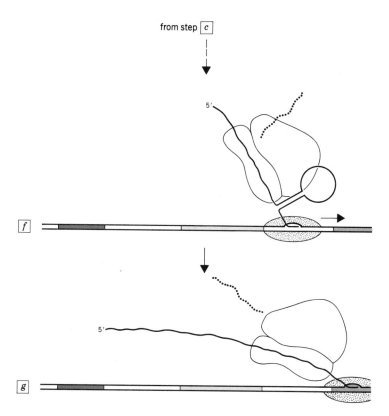

from step c

Figure 4-8 Attenuation. (a) In the upstream region of the tryptophan operon, the promoter is about 200 nucleotides from the beginning of the coding region of the first gene in the operon. **(b)** RNA polymerase binds to the promoter. **(c)** Transcription of the leader mRNA begins, ribosomes bind to the mRNA, and synthesis of the leader peptide begins. **(d)** In the presence of tryptophan, the ribosome moves along the leader RNA and a loop in RNA forms that acts as a transcription terminator. **(e)** Transcription stops, RNA polymerase is released from the DNA, and the mRNA is released from the polymerase–DNA complex. **(f)** In the absence of tryptophan, the ribosome stalls at a pair of tryptophan codons because there is no tryptophan to incorporate into the leader peptide. This allows a specific loop to form in the RNA that prevents the formation of the terminator loop. **(g)** RNA polymerase continues to travel down the DNA, synthesizing messenger. The first gene of the operon has its own ribosome binding site, so new ribosomes can bind to translate the message from the gene.

77

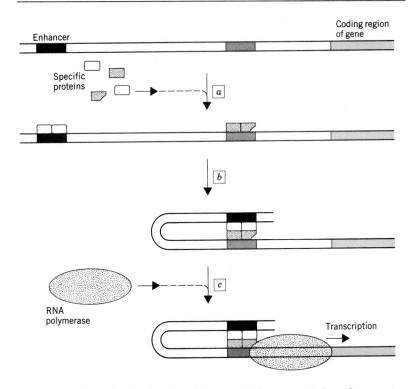

Figure 4-9 Gene Activation Involving an Enhancer. (a) Specific proteins bind to regions upstream from the gene. **(b)** The two sets of proteins bind to form a complex. **(c)** RNA polymerase binds to the complex.

gions of DNA near the gene. In these cases the DNA must loop to allow the various proteins to bind to each other and stimulate RNA polymerase action. A scheme for enhancer action is shown in Figure 4-9.

PERSPECTIVE AND REVIEW

During the past four decades we have learned that a remarkable molecular continuity exists among organisms. Many of the molecules of cells are constantly being converted into other molecules through a series of steps we call **metabolism.** These chemical conversions are stimulated by protein molecules called enzymes. Usually a different enzyme facilitates each step in a conversion pathway. To alter the characteristics of an organism—that is, to change the chemical reactions—one must change the proteins controlling these reactions. In principle, the protein content of an organism can be changed temporarily, as in the case of insulin injection by diabetics, or permanently, as with gene therapy. In all organisms on Earth the information required to make each protein is stored in a long, chainlike molecule called DNA. This information is arranged as short regions called genes, and each gene contains the information for the production of one protein. Genetic engineers change a particular protein by changing the information in DNA. Once the change in DNA has been made, every protein made from the new information will be of the new variety. The change is permanent because DNA is accurately reproduced every time a cell divides. By changing the information in DNA, humankind can change its own nature.

Within the framework just sketched, RNA seems to play the role of intermediary, acting as a messenger, an adapter, and a workbench. This may not always have been the case. We now know that some RNA molecules can speed up chemical reactions, just as proteins do. We also know that other RNA molecules can store genetic information, just like DNA (this is the case with some viruses). There is an emerging thought that at the beginning of life RNA molecules performed all the tasks that are now relegated to protein and DNA. That was the RNA world. Gradually RNA lost its supremacy because it fails to do the job as well as DNA and proteins. Proteins, since they are made of 20 different types of subunit instead of 4, can form a wider variety of structures than RNA; proteins can be more specialized. DNA is more stable than RNA. As a repository of information, DNA is the clear winner.

As cells evolved and gained the ability to live together in communities, fundamental aspects of replication, transcription, and translation remained the same. But multicellular life called for more genes and more elaborate control networks. We see the results of these changes when we compare prokaryotic and eukaryotic organisms. The eukaryotic cell is larger, and its functions are much more compartmentalized. Its proteins are often glycosylated to help them reach the correct cellular location safely. And its DNA is organized into many linear chromosomes that reside inside a nucleus rather than existing as a single circular chromosome unbounded by a membrane. Compartmentalization has required a number of adjustments. For example, eukaryotic messenger RNA must travel to the cytoplasm to be translated. To aid that journey, the molecules are often bound by protein and always have a protective cap at their 5' end. Most also are polyadenylated—that is, they have a long tail composed only of As (polyA). Other differences between prokaryotic and eukaryotic cells will emerge in later chapters, and each one poses special problems for the engineering of complex organisms.

Questions for Discussion

1. Describe how information flows, in terms of molecules, when gene expression occurs.
2. Although it seems most efficient to control gene expression by regulating when RNA synthesis occurs, sometimes control is exerted by regulating when translation occurs. In frog eggs, for example, translation control dominates: messenger RNA accumulates within the egg cell, and translation is delayed until fertilization by a sperm has occurred. How might this accumulation of RNA benefit the developing frog embryo?
3. Cases have been found in which a single type of repressor binds to operators of many genes that are widely separated on a chromosome. Inactivation of the repressor then leads to transcription from the whole set of genes. If the repressor gene is very active, removal of the inducer will then lead to rapid repression of all the genes in the system. For this system to work, the repressor must be kept at a concentration low enough to permit the inducer to easily bind most of the repressor molecules. How might repressor levels be controlled?
4. RNA polymerase is composed of several separate proteins called subunits. In bacteria one of these subunits is called **sigma.** Sigma is thought to be involved in the recognition of promoters: the preference of RNA polymerase for promoters changes when the sigma subunit associated with the enzyme changes. By

controlling the production of certain sigma subunits, it is possible to control the expression of large sets of genes. Under what conditions would this sigma subunit mode of regulation be more suitable than the repressor system described in question 3? Among the factors to consider are response time, ability to amplify the effect of the stimulus, and reversibility.

5. Occasionally a point mutation will occur in a gene, rendering its protein product nonfunctional (see Figure 3-5). At times, a second mutation, located in another gene, causes the cell to produce a normal, functional protein from the first gene. This second mutation is called a suppressor mutation. Often suppressor mutations fall in the anticodon region of transfer RNA genes. Using Figure 2-6, determine which nucleotide change or changes in the anticodon of the tRNA would suppress the mutation illustrated in Figure 3-5. Would you expect such a tRNA-type suppressor mutation to affect mutations in many genes, or would it be gene specific?

MANIPULATING DNA

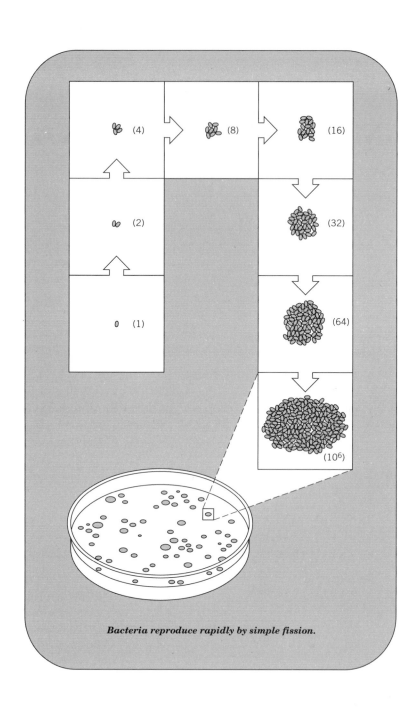

Bacteria reproduce rapidly by simple fission.

MICROBIAL GROWTH

One-Celled Organisms as Tools for Gene Cloning

Overview

Bacteria and yeasts are microscopic organisms that reproduce very rapidly and are easily grown in the laboratory. A single cell placed on a solid surface containing nutrients can multiply to form a colony comprising millions of identical cells. If the colony is transferred to liquid growth medium, the cells will continue to multiply; within a day or two, a culture containing trillions of identical cells can be harvested.

Single-celled microorganisms are used in two ways for genetic engineering. First, they help us isolate particular recombinant DNA fragments. Conditions can be obtained in which a small piece of DNA will pass into the interior of a cell where the DNA can replicate. Then the cell containing the piece of DNA can be separated from all other cells and allowed to reproduce until it forms a visible colony. The colony can be grown to provide large amounts of the recombinant DNA. Second, pure cultures, in which each cell contains the same type of recombinant DNA, can be grown in huge quantities to produce desired protein products.

INTRODUCTION

In Chapter 1 it was pointed out that organisms can be divided into two general types based on molecular and cytological (microscopic) characteristics of their cells. One group, the prokaryotes, are distinguished by lacking a nucleus for storage of DNA. The bacteria, which are tiny, single-celled organisms, are the principal members of this group. The other group, the eukaryotes, have a true nucleus. The eukaryotes include most other organisms, including ourselves (there is an intermediate group called archaebacteria, but a discussion of these organisms is outside the scope of this book). Compared to humans, bacteria are very simple organisms, and that simplicity has allowed us to gain a good understanding of their molecules. Consequently, features of bacterial life are often the best examples for making points about genes. Bacteria also serve an important function in gene cloning—after we prepare large collections of DNA fragments, bacteria can be used to help us fish out particular fragments. These two considerations make it important to understand the growth properties of bacteria that allow them to be so easily manipulated. Many of these properties are shared by single-celled eukaryotic organisms called yeasts. The lives of the yeasts are somewhat more complex than those of bacteria, and the study of yeasts has given us an understanding of our own cells that could not have come from bacterial studies. In addition, yeasts have become useful in gene cloning for cases in which bacteria do not produce properly modified proteins or when it is necessary to clone very large pieces of DNA. The following sections focus on how single-celled organisms are handled and how very tiny things can be "seen" without use of microscopes.

CHARACTERISTICS OF BACTERIAL GROWTH

Bacteria live almost everywhere: in the soil, on our skin, and in our intestines. When most people think about bacteria, diseases come to mind, diseases such as tuberculosis, plague, botulism, anthrax, cholera, and typhoid fever. Fortunately, most bacteria are not pathogenic (disease-causing). Indeed, some of the nonpathogenic species have become the favorite experimental subjects of molecular biologists. About 50 years ago, molecular biologists developed an interest in bacteria because these organisms are very tiny (a high power microscope is needed to see individual bacteria) and because bacteria seem to lead such uncomplicated lives. They simply grow and then divide in half to produce two new cells. Each of the new, daughter cells then expands until it, too, divides to form two more cells. Thus bacteria reproduce by simple **fission.** This simplicity produced the hope that everything about bacterial life could be understood in molecular terms, that the essence of life itself would be revealed.

Two properties of bacteria are particularly important for genetic engineering. First, each bacterium is only a single cell, lacking limbs, organs, and complicated developmental stages. Consequently, it is relatively easy to obtain large numbers of *identical* cells for the study of cell chemistry. Second, bacteria grow and multiply rapidly. Experiments that would take years with other organisms can be done in a single day with bacteria. For example, in the laboratory bacteria of many types divide every 40 minutes. Thus, during an 8-hour day, a batch of reproducing bacteria will go through 12 generations, greatly facilitating the study of how traits are passed from generation to generation.

It is easy to cultivate large numbers of bacteria because of the two properties just mentioned, small size and rapid growth. The recipe for growing bacteria is simple: place a few bacteria in lukewarm **broth,** and let the bacteria do the rest. They grow and divide, producing what is called a **bacterial culture.** Within 24 hours even a small flask of broth may contain billions of bacteria. Thus astronomical numbers of bacteria are readily obtainable.

Although the growth properties of many bacteria make them suitable tools for gene cloning, a species called *Escherichia coli* (**E. coli** for short) is used most extensively. *E. coli* is not particularly distinctive: it is a small rod-shaped organism (Figure 5-1) that is normally a harmless inhabitant of the human digestive tract. Like most other bacteria, its shape is maintained by a rigid coating called

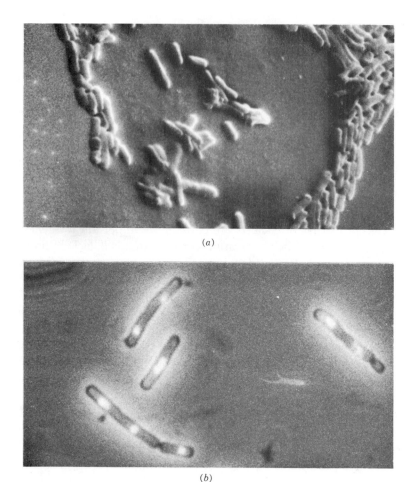

(a)

(b)

Figure 5-1 Photomicrographs of Bacteria. (a) A cluster of *E. coli* cells as they appear using scanning electron microscopy. Magnification is 3800 times. (Photomicrograph courtesy of Sandra McCormack, Rochester Institute of Technology.) **(b)** Several *E. coli* cells as they appear using light microscopy. The bright structures in the center of the cells are DNA-containing bodies called **nucleoids,** which have been stained with a dye called ethidium bromide. (An electron micrograph of a nucleoid that has been removed from the cell is shown in Figure 1-1.) The cells shown here are mutants that are unable to reproduce properly. As a result, the cells become elongated and appear to have more than one nucleoid per cell. Magnification is 3800 times. (Photomicrograph courtesy of Todd Steck, University of Rochester.)

a **cell wall.** Inside this wall the chemical reactions of life take place. *E. coli* is special because long ago a group of molecular biologists focused their research on it. As details about the chemistry of *E. coli's* life were learned, it became easier to conduct more sophisticated experiments on this bacterium than to start over with a new organism. As a result of this intense effort, *E. coli* has become the best understood organism on earth. Likewise, the best known cloning vehicles are infectious agents that use *E. coli* as a **host,** for they too have been intensively studied for many years.

Since our host for cloning, *E. coli*, is so small, special methods are required to use it. One cannot simply look through a microscope to see whether a particular cell has taken up a specific piece of DNA; the cells have few distinctive characteristics when viewed through an ordinary microscope. Only under special conditions can the DNA be seen inside the cell, and then it looks like a featureless blob (Figure 5-1*b*). Nor can one dissect a bacterium, as many of us dissected frogs in general biology class. Instead, indirect observations must substitute for what cannot be seen. For example, a liquid bacterial culture will appear cloudy if it contains more than 10 million cells per cubic centimeter (a quarter-teaspoon). Measuring the cloudiness (**turbidity**) is fairly easy because bacteria block the passage of light. If a test tube containing a culture of bacteria is placed in a light beam directed at a photodetector, the signal picked up by the detector is inversely related to the concentration of cells. Thus measuring the growth of a culture is also straightforward: as the bacteria multiply, the turbidity of the culture increases, and this effect is seen as a decrease in the amount of light that will pass through the culture.

BACTERIAL COLONIES

A single bacterial cell growing on a solid surface will multiply to form a cluster of cells called a **colony.** Since this is one of the more important concepts in gene cloning, a procedure is presented for obtaining bacterial colonies, a procedure that could be carried out in almost any kitchen. Biologists use slightly more refined equipment to obtain colonies, but the principles are the same. First **dissolve** in boiling water some **agar,** which is a gelatinlike substance (the process is similar to preparing gelatin dessert). Add sugar, minerals, and perhaps a rich source of nutrients such as beef extract. Pour the resulting solution into a **sterile,** covered dish and set aside, allowing the agar to cool and solidify.

Next place a teaspoon of soil in a tablespoon of water. Stir briefly. Soil is a good place to obtain microorganisms, since many types of bacteria live there. Bacteria such as *E. coli* can be obtained from sewage. Then find a thin piece of wire and bend the end to form a loop about a millimeter in diameter. Heat the loop with a flame to sterilize it. Dip the loop directly in the mixture of soil and water and lift it out. The loop will trap a tiny drop of water containing bacteria. Place the drop of water from the loop onto the surface of the solid agar, and smear the drop over the surface of the agar with the wire loop. Replace the lid on the dish, and allow the bacteria to grow and multiply at room temperature. Several days later, hundreds of small bumps will be seen on the surface of the agar. These bumps, which often look like glistening blisters about a millimeter across (see Figure 5-2 and frontispiece for Chapter 5), are bacterial colonies. When the drop of water was smeared over the agar, a small number of bacteria from the soil were scattered to widely separated spots on the agar. Each cell divided many times, and since the new cells could not move away from each other, they piled up. Within a day or so, the bacterial colony became visible. With this simple procedure it is easy to separate and culture the individual bacterial cells in the original soil sample. *The millions of bacteria in each colony all arose from a single bacterial cell.* Thus all cells in a colony are identical; they are members of a **clone.**

The ability to obtain individual colonies is important because persons who perform **gene cloning** cannot visually distinguish one gene from another. We cannot simply use forceps to pick particular genes out of a pile of DNA fragments. Instead, we use bacteria. Bacteria can incorporate small pieces of DNA if the DNA is linked to a cloning vehicle, a small, **infectious** DNA molecule. Thus gene cloners isolate DNA fragments by first transferring them into bacteria and then separating the bacteria from each other by spreading

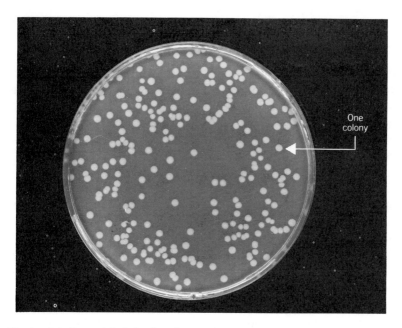

Figure 5-2 Bacterial Colonies Growing on Agar. A dilute suspension of *E. coli* cells was spread on solid agar in a petri dish and was incubated at 37°C. After 24 hours, colonies, each about 2 millimeters in diameter, became visible on the agar.

the broth containing them on agar. The bacteria grow into visible colonies, and the cloned DNA fragments multiply millions of times. The colonies are then tested for the presence of particular DNA fragments as described in later chapters.

Finding the colony that contains a particular gene is not easy; the gene cloner faces a problem of numbers. First, the gene being sought may be rare; it is common for the DNA fragment containing a particular gene to represent less than one out of 100,000 DNA fragments. Second, transferring DNA into bacterial cells is an inefficient process. In some cases fewer than one in 10,000 cloning vehicles will take up residency in a cell, and only a fraction of these will be attached to one of the fragments being sought. Thus the chance that any particular bacterial cell will contain a specific fragment may be less than one in a billion. The cloner's main task is to find that rare cell.

Whenever biologists work with an organism, they try to obtain a pure culture, one that contains only the type of organism being studied. Interpreting experimental results or producing a pure product is very difficult if other types of organism contaminate the culture. Since it is easy to grow single bacterial cells into colonies free from other organisms, obtaining pure cultures is straightforward.

The principle of pure cultures can be illustrated by describing one way to clone the human insulin gene using bacteria. First, human DNA is obtained (a method was presented in Chapter 2). The DNA is cut into millions of discrete fragments and inserted into cloning vehicles, producing many different types of recombinant DNA molecule. Cutting DNA and the use of cloning vehicles are described later chapters. The collection of recombinant DNA molecules is next mixed with a huge number of *E. coli* cells. Some of the recombinant DNA molecules get inside bacterial cells, and the cells are spread out on the surfaces of agar plates to separate one cell from another. The cloning vehicles usually contain a gene for antibiotic resistance, so addition of the appropriate antibiotic to the agar allows growth only of cells that contain the cloning vehicle. After colonies have arisen, each is biochemically tested by strategies described in later chapters until a colony containing insulin genes is found. Then that particular colony is carefully touched with a piece of thin, sterile wire, causing some of the bacterial cells in the colony to stick to the wire. The captured cells are then transferred into a flask of sterile broth by simply dipping the end of the wire into the broth. Some of the cells fall off the wire and begin to grow and divide in the broth. By the next day the flask will be full of bacterial cells, each of which has arisen from the single bacterium carrying the insulin gene. The flask contains what microbiologists call a pure culture of bacteria, one having only a single type of organism; all the cells in the flask are members of a clone. This culture can be maintained indefinitely if care is taken to keep other bacteria out of the culture. To grow large amounts of bacteria, one simply transfers a drop from the pure culture into a vat of sterile broth. Within a few days the vat will contain trillions of bacteria, each containing the insulin gene.

Storing and shipping bacterial strains is also straightforward. Often glycerol (glycerin) is added to a culture in a small vial, and the vial is stored in a freezer. At −80°C bacterial cells survive for decades. For shipping, a few drops of culture are spotted onto a small piece of sterile, absorbent paper. The paper is then wrapped in sterile aluminum foil and placed in the mail (precautions are required for harmful species).

ANTIBIOTIC RESISTANCE

Antibiotics are molecules, often produced by microorganisms, that either kill or block the growth of other microorganisms. We usually think about antibiotics in terms of the intracellular processes they attack. For example, **penicillin** and its relatives interfere with the construction of bacterial cell walls. Tetracycline, streptomycin, and erythromycin attack the bacterial machinery that makes new proteins. A few other antibiotics, such as rifampicin, have RNA polymerase as their target. A recent addition to our repertoire is ciprofloxacin. It attacks DNA gyrase, the bacterial enzyme that puts twists into DNA (Figure 2-7). In general, antibiotics bind to specific molecular targets and inactivate them.

Bacterial cultures become resistant to antibiotics through changes in (1) the antibiotic target that renders it resistant, (2) the cell wall of the bacterium that prevents entry of the antibiotic, (3) a pump that forces the antibiotic out of the cell, or (4) a detoxifying system that breaks the antibiotic down to a harmless substance. These changes can arise spontaneously. Bacterial cultures frequently contain huge numbers of cells (10–1000 million per milliliter), and random mutations arise often enough for a few members of the culture to be resistant. Addition of antibiotic to the culture allows resistant bacteria to overgrow the susceptible ones. Virtually all members of the population eventually become resistant.

In the late 1960s biologists discovered that bacterial cells are occasionally invaded by foreign DNA molecules (**plasmids**) that carry genes conferring resistance to one or more antibiotics. Methods were developed to isolate plasmids and introduce them into other cells. Since plasmids each carry an origin of replication, they replicate inside bacteria. The **antibiotic resistance gene(s)** on the plasmid makes it easy to find the few bacterial cells that took up the foreign DNA: they are the only cells to grow into colonies on agar containing the relevant antibiotic. Plasmids that carry antibiotic resistance are among the vehicles used to transfer cloned genes into bacterial cells. They are described in more detail in the next chapter.

YEAST CELLS

Yeast is a generic term applied to fungi that grow as single cells. These microorganisms are often found on plants, and some cause infections in humans. The most commonly studied yeast, *Saccharomyces cerevisiae,* is an agent of alcoholic fermentation. A typical yeast consists of small oval cells that multiply by forming a bud, a tiny protrusion from the cell that gradually enlarges until it is the same size as the mother cell. At that point the nucleus divides, and a cross wall forms between the cells, which then separate. Yeasts also undergo a form of sexual reproduction, and many of their biochemical processes are quite similar to those found in humans (studies with yeast led to the identification of a human gene involved in a form of cancer). Since yeasts grow into colonies on agar, they are handled like bacteria. That makes them useful for cases in which bacterial features are incompatible with proper activity of the protein produced by a given cloned gene. In addition, methods have been developed with yeast in which very large pieces of DNA can be cloned into artificial chromosomes (YACs) for sequence determination.

PERSPECTIVE

Throughout our history bacteria have plagued us by causing disease. We have learned to live with these tiny organisms by

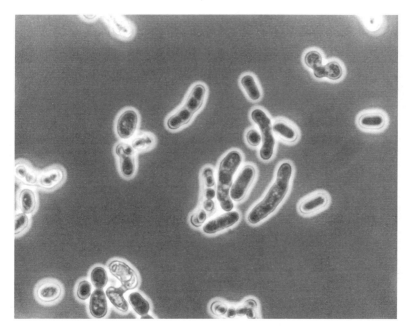

Figure 5-3 Yeast Cells. (Photomicrograph courtesy of Dr. Eric Chang, New York University.)

washing our hands, treating our drinking water, immunizing our bodies, and keeping bacteria-infected fleas and ticks from biting us. Occasionally our efforts fail, and we have to kill bacteria with antibiotics, an ability we are rapidly losing as resistant forms emerge. We have also learned to use bacteria. For example, bacteria help us make products such as yogurt. In the early 1940s we began to use bacteria to study the molecules of life, and we discovered that there is an amazing continuity among all life forms when viewed at the molecular level. The intense study of bacteria led directly to gene cloning: all the tools are of bacterial origin. Now we are using these tiny organisms to help understand and manipulate our own chemistry.

Yeast cells grow like bacteria, and so mutations can be studied effectively. Yeast molecules act more like ours than do those of bacteria, so yeasts have been particularly useful for understanding the reproductive cycle of eukaryotic cells. The experience gained from these studies with yeast led naturally to the use of these micro-organisms as hosts for foreign DNA molecules during the gene cloning process.

One of the simplest manipulations of single-celled organisms, the growth of cells into visible colonies, turns out to be one of the most important for gene cloning. A brief consideration of antibiotic susceptibility illustrates why. For *E. coli,* hundreds of millions of cells can be spread on the surface of an agar plate, and no growth will appear if all the cells are sensitive to the antibiotic. But if a single resistant cell is present, it will multiply to form an easily seen colony. Thus growth of colonies from single cells allows us to fish out very rare members of a population.

Questions for Discussion

1. Bacteria are about 1 to 2 **micrometers** (μm) long. Compare their size to other small objects with which you are familiar.
2. In broth containing many nutrients, some bacteria grow rapidly, dividing every 20 minutes. At this rate, how many cells would there be in a culture after 10 hours of growth if the culture started with only one cell?
3. A bacterial colony contains roughly 1 million to 10 million cells. Describe how you would determine the number of live cells in a colony growing on an agar plate using only a spatula to scrape off the colony, test tubes, liquid growth media, pipettes to measure liquid volumes, agar plates, and an incubator.
4. The chemical structure of DNA is changed (damaged) by radiation (X-rays, ultraviolet light, etc.). Sometimes the radiation can even break DNA. How do you think DNA breaks would affect

bacterial growth? How do you think DNA damage arising from sunlight might affect your skin?

5. Design a way to measure the amount of DNA dissolved in water using the knowledge that DNA absorbs ultraviolet light. (Hint: How can you use light to measure the number of bacterial cells suspended in water?)

Resuspension buffer - EDTA - lyses bacterial cell
  1. overall structure of cell envelope lost.
  2. Some cellular enzymes that degrade DNA inhibited.

Alkaline - SDS   disrupts H bonds.
  increases pH. 12 - 12.5   2 strands of DNA
                                          separated
                                                  - denatured
3M Sodium Acetate ↓ pH 4.8.

concentrate plasmid DNA
  - ETHANOL PRECIPITATION. centrifuge
      pellet contains wanted DNA.

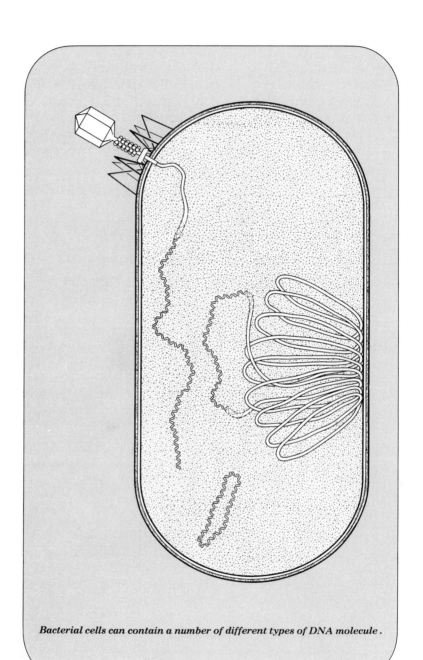

Bacterial cells can contain a number of different types of DNA molecule.

C H A P T E R   S I X

PLASMIDS AND PHAGES

Submicroscopic Parasites Used to Deliver
Genes to Cells

Overview _____

Bacterial plasmids and phages are minute agents that infect bacterial
cells and then use the bacterial components to replicate themselves. They
contain genetic information, which in most cases is stored in short DNA
molecules. Molecular biologists have found plasmid and phage DNA
molecules particularly easy to handle and study. Genetic information
from animals and other organisms having very long DNA molecules is
difficult to manipulate unless the DNA has been cut into small pieces
and the pieces have been physically separated. Plasmids and phages
assist in separating the DNA pieces. Since DNA fragments can be inserted
into plasmid or phage DNA without impairing infectivity, these tiny
infectious agents can be used as vehicles to place almost any DNA
fragment inside a living bacterial cell. There the DNA fragment reproduces
as a part of the plasmid or phage DNA.

INTRODUCTION

Certain types of small DNA molecule are infectious. Once inside a living cell, they can utilize the machinery of the cell to reproduce. Infectious DNAs can have profound effects on living cells. Some types take over a cell and kill it, while other types can be beneficial to the host. Infectious DNA molecules fall into two general types, the plasmids and the viruses. Plasmids are naked DNA molecules; they are generally circular and are found only inside cells. Viruses surround their DNA molecules with a protective shell of protein; thus, they can sometimes survive for many years outside a host cell. A virus that infects a bacterium is called a **bacteriophage** or simply a **phage.**

It is possible to cut plasmid and bacteriophage DNA in a specific place, insert a piece of DNA from a different source, and still retain all the information necessary for infection and replication by the plasmid or phage. Thus these infectious DNA molecules are useful as tools to transfer DNA from one type of cell to another; they are the cloning vehicles referred to in earlier chapters.

Before delving into what happens when DNA molecules invade cells, it is useful to restate the problem that cloning vehicles help overcome: the specific fragment of DNA one wishes to obtain must be separated from the thousands (or sometimes millions) of other fragments produced during the cutting and joining process, and then the fragment must be located. The separation aspect is not a problem per se; one could place a drop of water containing DNA fragments on an agar plate, smear the drop over the whole surface of the plate, and separate the fragments easily enough. But then it would be very difficult to detect the fragments. They are so small that they are visible only with the aid of an **electron microscope,** and this instrument is impractical for scanning a large surface or for distinguishing one stretch of DNA from another. Moreover, DNA fragments spread on an agar plate would be too dilute to find with a complementary radioactive probe (described in Chapter 8). In one sense, such a spreading process would be much like scattering straw from a haystack over a large field. Finding one particular straw would be next to impossible. If the straw of interest had seeds attached, however, we could simply wait until the seeds sprouted into a plant; the plant could be seen, and the particular straw would be found at the base of the plant.

In the case of DNA fragments, one needs a way to multiply the fragments after scattering them. Then radioactive probes can be used to find specific fragments. Gene cloners multiply the fragments using cloning vehicles and bacterial cells (cloning vehicles are also available for placing DNA fragments in yeast and animal cells). First, the vehicles and their attached DNA fragments are transferred into bacterial cells (no more than one vehicle and fragment per cell). Next, the cells are scattered on the surface of an agar plate. Third, the cells are allowed to multiply. The cloning vehicles, along with the attached DNA fragments, also multiply.

It might appear that cloning vehicles are unnecessary in the process just described. One need only get the DNA fragments into bacterial cells so that each cell obtains but a single fragment; then the cells can be spread out on an agar surface. Each cell will multiply to form a colony, and all the colonies can be tested for the gene of interest. Indeed, many types of bacteria will ingest DNA through their cell walls, but very few DNA fragments have the necessary start and stop signals to cause the cellular machinery in bacteria to replicate the DNA fragment. If no DNA replication occurs, the fragment will be diluted out as the bacteria grow and divide, for only the original bacterial cell will contain the DNA fragment. Even after many cell divisions, only one cell in the colony will contain the fragment. To bypass this problem, DNA fragments are first inserted into plasmid or phage DNAs, which contain the correct signals for replication. The plasmid and phage DNAs used as cloning vehicles then enter the bacterial cells and multiply as the cells multiply.

PLASMIDS

Most bacterial plasmids studied to date are small, circular, double-stranded DNA molecules (Figure 6-1) that occur naturally in their hosts. Like all natural DNA molecules, plasmids contain a special region in their DNA called an origin of replication. As pointed

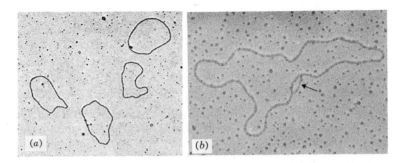

Figure 6-1 Electron Micrographs of Plasmids. (a) Four DNA molecules of the type called Col E1. These small, circular DNAs are only 0.001 times the length of *E. coli* DNA (compare with Figure 1-1). (Photomicrograph courtesy of Grace Wever, Eastman Kodak.) **(b)** Enlargement of a plasmid similar to Col E1. Sample preparation conditions were adjusted so that a short region of DNA would become single-stranded (arrow). (Photomicrograph courtesy of G. Glikin, G. Gargiulo, L. Rena-Descalzi, and A. Worcel, University of Rochester.)

out in Chapter 3, the origin serves as a start signal for DNA polymerase and ensures that the plasmid DNA molecule will be replicated by the host cell. Plasmids differ in length and in the genes contained in their DNA. Some of the smaller plasmids, which are popular in gene cloning, have about 5000 nucleotide pairs, enough DNA to **encode** about five average-sized proteins. By comparison, *E. coli* contains slightly more than 4 million nucleotide pairs in its DNA, and the human genome has about 3 billion nucleotide pairs.

An important aspect of plasmid DNA molecules is that they often make their host bacterial cell resistant to antibiotics, as pointed out in Chapter 5. Antibiotic resistance turns out to be extremely useful in genetic engineering. Consider, for example, DNA fragments inserted into a plasmid DNA having a gene that confers resistance to penicillin. When the plasmid DNA is added to a concentrated culture of penicillin-sensitive bacteria, plasmid DNA enters the bacterial cells and multiplies along with the cells. Next, the bacteria are spread on an agar plate containing penicillin and incubated overnight. That kills most of the cells (few actually acquire a plasmid when treated with DNA). The few cells that do obtain a plasmid are penicillin resistant, and they form colonies. Every colony growing on the agar plate contains cells harboring copies of the plasmid. Thus, when testing colonies for specific genes, gene cloners use antibiotics to avoid examining the millions of bacterial colonies that *fail* to take up a plasmid.

The process of causing bacterial cells to take up and replicate DNA is called **bacterial transformation** because the bacterial cells are converted from one type to another. In the example used, penicillin-sensitive cells were transformed into penicillin-resistant cells. Often bacteria must be treated in special ways before they will take up plasmid DNA. In the case of E. *coli*, the cells are first suspended in a dilute solution of calcium chloride that has been chilled. When cells are ready to receive plasmid DNA, they are said to have **competence.** For some bacteria, taking up DNA from the environment is a natural part of their lives, and they have a set of genes that is involved in creating the competent state.

In addition to the plasmids used for cloning, which are relatively small DNA molecules, there exist large plasmids whose protein products allow them to naturally go from one bacterial cell to another. The process of plasmid movement is called **conjugation.** The best studied case of conjugation involves a plasmid of E. *coli* called **F** or **fertility factor.** The plasmid DNA encodes proteins responsible for the formation of long, filamentlike structures (**pili**) that protrude from the outside of the bacterial cell. Pili are thought to be important in attaching or attracting plasmid-containing cells to plasmid-deficient ones. When the two cell types get close enough, they mate. One strand of the plasmid DNA is transferred from the plasmid-containing cell to the plasmid-deficient one. As this occurs, both strands are used as templates to synthesize new DNA. Thus both cells of the mating pair end up infected with a plasmid (Figure 6-2). On rare occasions the plasmid DNA and the bacterial chromosomal DNA join to form a giant circle. When this happens, genes in the bacterial chromosome can be transferred from one cell to another by the plasmid mating process. In a sense, bacterial conjugation is a primitive form of sex.

In general, transmissible plasmids are not good cloning vehicles. They are difficult to handle because they are so large, and their

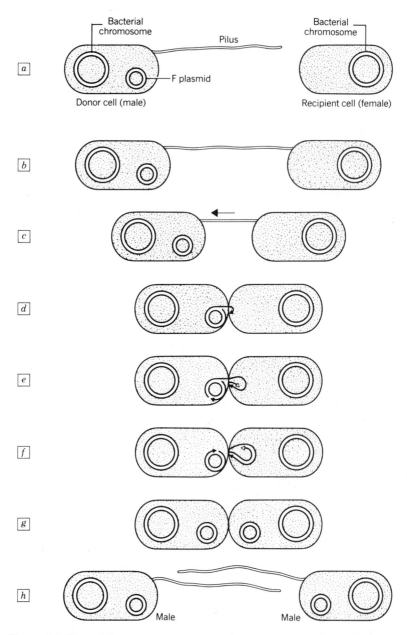

Figure 6-2 Bacterial Conjugation. (a) An *E. coli* cell containing an F plasmid forms long, flexible tubular structures (pili) on its surface (1–3 per cell). Each pilus is composed of proteins encoded by genes located on the F plasmid. **(b)** One pilus binds to an *E. coli* cell that lacks an F plasmid.

ability to move from one bacterium to another presents a potential hazard. For example, if transmissible plasmids were used to clone genes into laboratory bacteria, which are generally quite weak compared to wild strains, and if the "enginee red" bacteria containing the plasmids and cloned genes were to accidentally escape from the laboratory, the plasmids and the cloned genes could conceivably be transferred, through the process of conjugation, into a healthy, wild strain of the bacterium. The wild strain might have a greater chance than the weak, laboratory strain to enter human digestive tracts and thus might place an unwanted gene where it could potentially do harm.

OBTAINING PLASMID DNA

Before inserting DNA fragments into plasmids in preparation for cloning, a gene cloner must obtain a large amount of plasmid DNA. The first step is to prepare a liquid bacterial culture containing billions of cells harboring plasmids (see Chapter 5 for culture techniques). Then the DNA molecules must be removed from the cells. Bacterial cells that have grown in broth culture are first concentrated; one procedure is to place the broth in a test tube and allow the cells to gravitate to the bottom of the tube much like silt settling out of river water. Generally molecular biologists speed up the settling process by putting the test tube in a centrifuge as described in Figure 3-7. The bacteria are driven to the bottom of the tube, where they form a tight pellet. The bacteria remain

---

(c) The pilus retracts, pulling the two cells close to each other. (d) A break occurs in one of the DNA strands of the F plasmid; one of the ends of the broken strand rolls off the circular strand and passes into the recipient cell. (e, f) Soon after the single strand has reached the interior of the recipient cell, DNA polymerase (not shown) begins to make a complementary strand. At about the same time, a new copy of the transferred strand is made in the donor cell. (g) Both donor and recipient cells now have a complete copy of the F plasmid. (h) The two cells separate. Each contains a copy of the F plasmid. The recipient cells form pili. At this point both can act as donor cells.

compacted even after the tube has been removed from the centrifuge, so the broth in which they were growing is easily poured off. A small volume of water is then added to the trillion or so bacteria, and the tube is shaken to distribute the bacteria in the water. Enzymes and detergents are added to this suspension to dissolve the cell walls of the bacteria, releasing both bacterial DNA and plasmid DNA molecules from the cells. At this stage the content of the test tube is called a **cell lysate.**

Once the DNA molecules have been freed from the cells, the plasmid DNA must be physically separated from the bacterial DNA. The two types of DNA differ mainly in their length, with the bacterial DNA running as much as a thousand times longer, depending on the particular plasmid. Since bacterial DNA is so large, it tends to sediment to the bottom of a test tube faster than the small plasmid DNA. Consequently, one step in the purification procedure is to put the cell lysate in a tube that is spun in a centrifuge. The large bacterial DNA will form a pellet in the bottom of a test tube, while the much smaller plasmid DNA will stay in the upper fluid. Unfortunately, there is usually so much more bacterial DNA than plasmid DNA that this centrifugation procedure does not completely separate the two types of DNA molecule.

The great length of the bacterial DNA, however, makes it possible to carry out an additional type of separation. Bacterial DNA is easily broken by sucking the DNA-containing solution into a pipette. This procedure does not break the smaller plasmid DNA; after a few squirts through the pipette, bacterial DNA becomes linear while the plasmid DNA remains circular. Circular DNA has a distinctive property that allows it to be separated from linear DNA. This property is related to the concepts of buoyancy and relative density.

Consider for a moment your own buoyancy in water. If you are totally relaxed and motionless, you tend to float, with only the top of your head above the surface. You can change your buoyancy in two ways: by putting on a life jacket or by holding rocks in your hands. The life jacket lowers your overall buoyant density, causing you to float higher. The rocks increase it, causing you to sink lower. In addition to your own density, the density of the water is important in determining whether you sink or float. For example, water containing a high concentration of salt has a high density, and as a result people easily float in very salty water such as that of the Dead Sea.

Laboratory test tubes can be filled with salt solutions in such a way that the salt concentration gradually increases from top to bottom. The gradual change in salt concentration is called a concentration gradient, and it produces a **density gradient.** Molecules such as DNA will sink until they reach a solution density equal to their own density. At that point the DNA is at **equilibrium,** and if unperturbed, it will remain at that position forever. The important point to remember is that the depth to which a DNA molecule sinks in a density gradient depends on the density of the DNA and the density of the solution.

Now let's return to circular and linear DNA molecules. Both types consist of the same four bases, so physically they should have the same density. However, it is possible to add a dye molecule to DNA that acts as a life jacket, lowering the density of the DNA. Linear DNA molecules can bind more dye than circular ones: thus, linear molecules can effectively put on more life jackets, and they will not sink as far. Why do more dye molecules bind to linear DNA than to circular DNA? As the dye molecules bind to DNA, they insert themselves between the base pairs and slightly unwind the DNA (Figure 6-3*a*). Extreme examples of unwinding are shown in Figure 6-3*b* and *c* to illustrate the difference between linear and circular DNA. As DNA unwinds, it twists. Twists introduced into a linear DNA molecule are quickly lost as the free ends of the molecule rotate over each other (Figure 6-3*b*). In contrast, twists put into a circular molecule are retained because there are no free ends (Figure 6-3*c*). As more and more dye molecules bind to circular DNA, the twists accumulate, and it becomes increasingly difficult for dye molecules to bind. This is not the case with linear DNA. Thus the absence of ends for strand rotation prevents a circular DNA molecule from binding as much dye as a linear one.

The process is easier to carry out than to explain. A DNA preparation is mixed with the dye and a heavy salt in a test tube. At this stage one could rely on the force of gravity to generate the density gradient with the salt water, but it would take a very long time for the salt molecules to accumulate in the bottom of the tube. It would also take a long time for the DNA molecules to settle to their own density. The time required can be shortened by spinning the tube

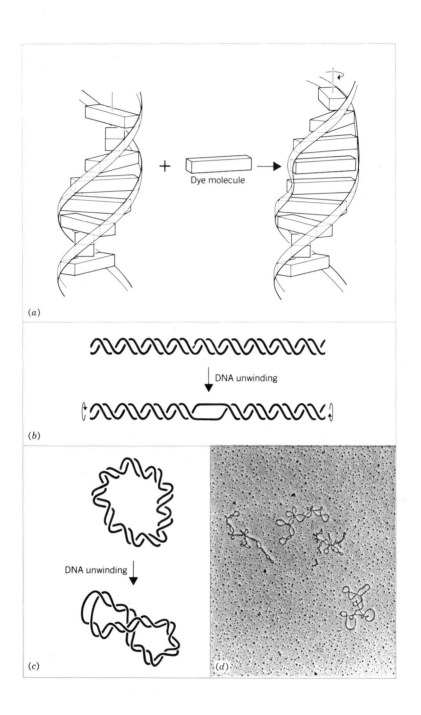

(a)

DNA unwinding

(b)

DNA unwinding

(c)

(d)

containing DNA, salt, and dye in a centrifuge. Afterward, the tube is examined with ultraviolet light (black light). When dye molecules bound to DNA absorb the ultraviolet light, they emit a visible, fluorescent light. Two bands can be observed in the tube. The upper one corresponds to the linear DNA (bacterial DNA) and the lower one to the circular plasmid DNA (see Figure 6-4). The circular plasmid DNA can be sucked out of the tube with a pipette, which will then contain pure plasmid DNA.

Now we can sketch how plasmids are used for cloning. Once the plasmid DNA molecules have been isolated, they are ready to be cut for insertion of a collection of DNA fragments. This is done by incubation with the appropriate restriction endonuclease (Chapter 7), followed by incubation with both DNA ligase and the collection of DNA fragments to be inserted. This produces a collection of recombinant DNA molecules, which is delivered to bacterial cells by mixing the cells and DNA under the proper conditions. Colonies are grown and then tested by nucleic acid hybridization (Chapter 8) to find those carrying the recombinant DNA of interest. At that point a large batch of recombinant-containing bacteria is prepared from a colony, and the plasmid DNA is isolated as described above. That DNA contains the fragment of interest.

Figure 6-3 Binding of Dye to Linear and Circular DNA Molecules. (a) Dyes such as ethidium bromide are flat molecules that resemble DNA base pairs. When such a dye binds to DNA, it slips in between two adjacent base pairs (shown as blocks). Although the hydrogen bonds holding the base pairs are not broken by the dye, the DNA double helix unwinds slightly (26° per dye molecule bound). **(b)** Unwinding twists the DNA, but the twisting of linear DNA dissipates as the ends of the strands rotate. For illustrative purposes, an extreme case of unwinding is shown, and base pairing has been disrupted. **(c)** No free ends are present in circular DNA, so twists arising from unwinding accumulate. The twists make further unwinding, and thus dye binding, more difficult. **(d)** Electron micrograph of twisted plasmid DNA molecules. (Photomicrograph courtesy of G. Glikin, G. Gargiulo, L. Rena- Descalzi, and A. Worcel, University of Rochester.)

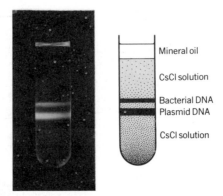

Figure 6-4 Separation of Plasmid and Bacterial DNAs by Dye Buoyant Density Centrifugation. Plasmid DNA, bacterial DNA, water, dye (ethidium bromide), and a heavy salt (cesium chloride: CsCl) were mixed in a plastic tube and centrifuged for 2 days at 35,000 rpm. Mineral oil, added to fill the plastic tube before centrifugation, prevents collapse of the receptacle from the force of the centrifugal field. After centrifugation, the tube was illuminated with ultraviolet light, and bright orange bands appeared in the tube, indicating the location of the DNA molecules.

BACTERIOPHAGES

Bacteriophages, commonly called phages, are viruses that infect bacteria, and they are more complicated than plasmids. In addition to having an origin of replication, phage DNA contains genes encoding proteins that form a protective coat around the DNA. But, like plasmids, phages lack the machinery necessary to actually make proteins; consequently, they reproduce only inside living bacterial cells. Both phages and plasmids can be used to separate and amplify specific DNA fragments, but the two have very different means of reproduction.

Some phages are like miniature hypodermic syringes (Figure 6-5). The phage DNA is wrapped into a tight ball inside a headlike structure made of proteins. A tail, also made of proteins, is attached to the head. When such a phage particle comes in contact with a bacterial cell, the phage tail sticks to the cell wall, and the DNA is squirted from the head, through the tail, and into

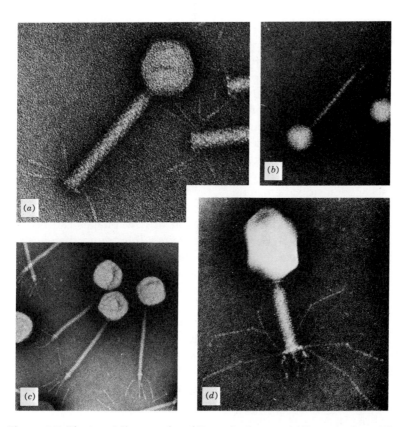

Figure 6-5 Electron Micrographs of Bacteriophages. (a) Bacteriophage P2, magnification 226,000 times. **(b)** Bacteriophage lambda, magnification 109,000 times. **(c)** Bacteriophage T5, magnification 91,000 times. **(d)** Bacteriophage T4, magnification 180,000 times. (Photomicrographs courtesy of Robley Williams, University of California, Berkeley.)

the bacterium (see frontispiece, Chapter 6). Soon after the phage DNA gets into the cell, it begins to take control. Special phage genes are transcribed by the bacterial RNA polymerase, and the resulting messenger RNAs are translated into phage proteins using the bacterial ribosomes. At early stages of infection some phages produce proteins that destroy the bacterial DNA, chopping it into individual nucleotides. Once that has happened, the bacterium is doomed, because all the information needed for its reproduction is gone. Some phages have genes that produce an RNA polymerase, so they do not have to rely on the host polymerase to make messenger RNA from phage genes late in infection.

Many phages also have genes for their own DNA replication machinery. When this apparatus is in place, the phages can use nucleotides released from the bacterial DNA to make phage DNA. Hundreds of copies of the phage DNA are made, and within minutes other genes on the phage are turned on to produce new head and tail proteins. The head proteins assemble into heads, phage DNA is packaged inside them, and a tail is attached to each head. The assembly of new phages occurs *spontaneously,* and the total time from injection of DNA to production of new phages can be less than 20 minutes. The bacterium becomes little more than a shell containing hundreds of new phage particles. As a final act, the phage produces an enzyme that destroys the bacterial cell wall, releasing the phage particles to seek new hosts.

Phage efficiency is awesome. One phage can produce hundreds of **progeny** particles. Each progeny particle can then infect a bacterial cell and produce hundreds more phage particles. By repeating the infection cycle just four times, a single phage particle can lead to the death of more than a billion bacterial cells.

Now consider what happens if a small number of phage particles is added to a *dense* bacterial culture, and before the first bacterial cell has been broken open by the newly made viruses, the culture is spread on an agar plate. Within 20 to 30 minutes the first infected bacteria rupture, releasing phage particles. So many bacteria are on the agar plate that the new viruses quickly attach to nearby bacteria and repeat the infection process. Gradually the circle of death expands away from the point at which the original infected cell had fallen on the agar surface. Meanwhile, the uninfected cells, which are the vast majority, continue to divide. They are unaffected by the fate that has befallen a few of their number, and as they become more numerous, they begin to deplete the food supply. Eventually, the uninfected bacteria completely cover the agar surface except for the small regions in which the phages are attacking. Within a day the bacteria stop growing, and their biochemical machinery shuts down. The phages, too, stop reproducing, since they rely on active bacterial machinery to supply energy for their enzymes. Thus the agar plate is covered by a lawn of bacteria containing small holes wherever the phages had been multiplying (Figure 6-6). These holes, called **plaques,** are about an eighth of an inch in diameter, and each one arose from a single phage particle.

If a DNA fragment is inserted into a phage DNA molecule without destroying important phage genes, the phage, when it infects a bacterial cell will reproduce the fragment along with its own DNA. Gene cloners can determine which plaque has a particular piece of DNA by using a radioactive probe to test the plaques for DNA having base pairs complementary to a specific gene (Chapter 8).

Once the right plaque has been found, a piece of sterile wire is poked into it. A small number of virus particles will attach to the wire. When the wire is stuck into a fresh culture of bacteria, the phages drop off the wire, attack the bacteria, and reproduce to form trillions of progeny. Large amounts of phage DNA will be made and packaged. Since the cloned gene is actually a part of the phage DNA, it too will be very abundant; moreover, it will be packaged inside a phage head along with the phage DNA sequences. Phage particles can be easily purified by density-gradient centrifugation in a way similar to that described earlier for purification of plasmid DNA. Since DNA and protein have different densities, phages, which are a combination of DNA and protein, will have a buoyant density between that of pure DNA and pure protein. Thus, no dyes are needed to separate phages from other cellular components, including bacterial DNA. A test tube containing the phage in a heavy salt solution is centrifuged until a density gradient is established, whereupon the phage sediments to its own density. When the tube is removed from the centrifuge and examined, the phage will appear as an opalescent band that can be easily sucked out with a pipette (Figure 6-7). The DNA of that phage contains the DNA fragment of interest.

One of the phages used for cloning is called **lambda** (Figure 6-5*b*). When lambda DNA is injected into a bacterial cell, it has two choices. It can behave as described above and destroy the bacterium. Or, it can take up residence in the cell and enter a

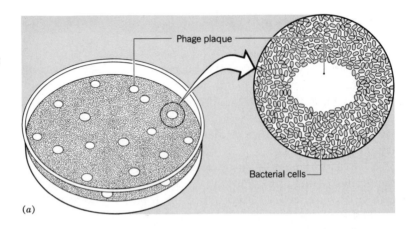

(a)

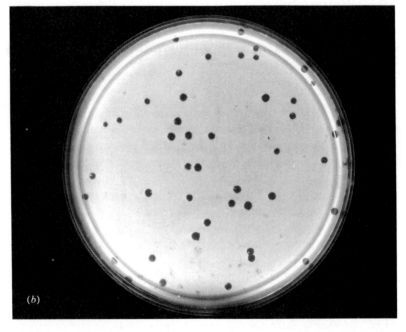

(b)

Figure 6-6 Bacteriophage Plaques. An agar plate is covered by a lawn of bacteria. The holes in the lawn, called plaques, are regions where phages have killed the bacteria. **(a)** Schematic diagram. **(b)** Photograph of agar plate. (Photograph courtesy of Robert Rothman, University of Rochester.)

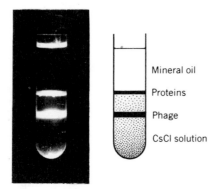

Figure 6-7 Purification of a Bacteriophage by Centrifugation. A bacterial lysate containing phage particles was placed in a plastic tube, mixed with cesium chloride, and centrifuged at 25,000 rpm for one day. During centrifugation the phage particles formed a band as indicated in the figure. The band above the phage is composed of bacterial proteins and cell wall material. As in Figure 6-4, the tube was filled with mineral oil to prevent collapse during centrifugation.

quiescent mode. When the latter choice is made, the lambda DNA inserts (integrates) into the bacterial chromosome; it becomes part of the bacterial DNA (Figure 6-8). Our knowledge of **integration,** a natural joining of DNA molecules, is important for understanding a number of important phenomena including the biology of AIDS (Chapter 12). Accompanying integration is a change in phage gene control. The phage genes that normally would produce the proteins to kill the cell are turned off by a repressor protein made from a phage gene. Thus every time the bacterial DNA replicates, lambda DNA replicates along with it. In this dormant state, lambda DNA does little more than produce repressor protein to keep its own genes shut down. At the same time, the repressor protects the bacterial cell from infection by other lambda phages— when the newcomers inject their DNA, it is quickly bound by a repressor produced from the resident lambda. All genes necessary for replication are shut off, and thus the incoming phage is unable to initiate a **lytic infection** that would kill the cell. Consequently,

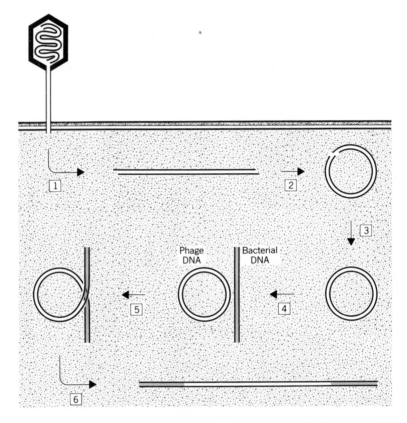

Figure 6-8 Formation of a Lysogen. **(1)** Bacteriophage lambda injects DNA through the bacterial cell wall; the resulting linear DNA molecule has sticky ends. **(2)** The DNA circularizes and **(3)** becomes ligated. At this point the phage has two choices. It can integrate its DNA into the bacterial DNA **(4–6)** and remain quiescent for an indefinite number of bacterial generations. The alternative (*not shown*) is to replicate its DNA, produce progeny phage, and kill the bacterium. Both phage and bacterial proteins play important roles in the integration process.

one can easily find **lysogens** (bacterial cells that are being protected by a resident phage) by looking for bacterial colonies in the middle of a phage plaque.

If the repressor that shuts down lambda gene expression is destroyed, as happens when cells are exposed to ultraviolet light, the phage DNA removes itself from the bacterial chromosome and directs the cell to make phage particles. The bacteria then become filled with phage particles, break open, and release phages into the environment.

PERSPECTIVE

Plasmids and phages are among the smallest and most efficient infectious agents in nature. Some are so efficient that they use the same nucleotide sequence to encode two different proteins. Plasmids have received considerable attention because of their medical importance: some plasmids carry genes that make their host bacteria resistant to antibiotics. Through our massive use of these drugs, we have encouraged the spread of plasmids to the point that almost every type of bacterium pathogenic to man now carries these infectious drug resistance factors.

Phages are important in another way. For five decades molecular biologists concentrated on discovering how phages regulate their genes and replicate their DNA. As a result, phage studies provide the basis for most of our understanding of DNA. As our studies shift to gene hunting in complex organisms, plasmids and phages are assuming a new role in biology, that of workhorses harnessed to purify genes.

More complex cells also contain small DNA molecules: some are considered to be plasmids; others are designated as viruses. Thus many of the principles discussed in this chapter apply to all cell types, and a variety of these small DNAs are now being used to manipulate genes in animal cells. Even the principle of lysogeny is important to us, for a version of that theme is used by the AIDS virus after it has infected human cells.

Questions for Discussion

1. What distinguishes a virus from a cell? Is a virus alive?
2. A bacteriophage can sometimes kill its host within 20 minutes, and in the course of infection 100 phage particles might be released from each cell killed. If you begin with a single phage particle in a culture of sensitive bacteria, and assuming you do not run out of bacteria, how many phage particles will you have in 2 hours?
3. Some bacterial viruses carry a gene for producing their own RNA polymerase. The promoters recognized by the viral polymerase differ from those used by the host RNA polymerase, thus providing the virus some control over its own destiny. If no RNA polymerase protein is carried in the virus particles, must the viral RNA polymerase gene have a promoter recognized by the host or viral RNA polymerase?
4. Devising ways to detect phages and plasmids is crucial to learning about them and to using them. Phages are generally detected by their ability to lyse bacteria, and their number is determined by counting plaques on a lawn of bacteria. Describe several ways to determine whether the cells in a bacterial colony contain a plasmid.
5. Some phages have RNA as their genetic material. How does that composition affect their ability to be cloning vehicles?
6. Figure 6-1 shows that single-stranded DNA can be distinguished from double-stranded DNA by electron microscopy. If two strands of a DNA molecule are not completely complementary when hybridized, they will not form a complete duplex, and the regions of noncomplementarity can be mapped by an electron-microscopic procedure called **heteroduplex mapping.** Looking ahead to Figure 7-3, you can see two plasmid molecules: the cloning vehicle and, ultimately, the recombinant DNA molecule. Suppose you cut both plasmids only once and at the same nucleotide sequence. This will produce linear DNA

molecules. You next mix the DNA molecules, heat them to separate the strands, and then incubate them at the appropriate temperature to allow hybrids to form. You then examine the DNA in the electron microscope. Draw pictures of the types of structure you expect to see.

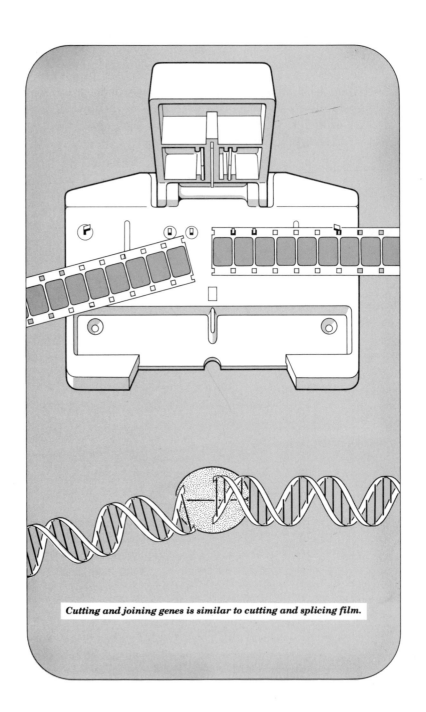

Cutting and joining genes is similar to cutting and splicing film.

CUTTING AND JOINING DNA

Restriction Endonucleases and Ligases Used to Restructure DNA Molecules

Overview

Gene cloners move specific bits of genetic information from one DNA molecule to another by cutting and joining procedures that utilize specific enzymes. DNA is very long and contains a large number of sites at which cutting can occur. Consequently, cutting often results in the creation of many different fragments in terms of length and nucleotide sequence. Biochemical methods based on the principle of complementary base pairing are used to detect and locate *specific* fragments after they have been joined to small infectious DNA and delivered to microorganisms.

The enzymes that cut DNA are called **restriction endonucleases.** Specific cutting produces DNA fragments having discrete lengths. Since nucleases can be obtained that recognize different sites in a DNA molecule, it's possible to cut DNA at a wide variety of locations. Cuts by different enzyme types produce DNA fragments having different sizes; two enzymes cutting the same DNA at the same time will cut in different places, producing smaller fragments than will result from either enzyme cutting alone. It is possible to locate the cutting sites of one enzyme relative to another by comparing the sizes of the DNA fragments produced by the respective enzymes. This type of analysis produces a restriction map characteristic of the particular DNA being studied.

A mutation occurring in a DNA molecule at a site normally cut by a restriction endonuclease will prevent the enzyme from attacking at that point; consequently, a larger DNA fragment will be produced when mutant DNA is cut. Other mutations can add restriction sites, making fragments smaller. Addition or deletion of DNA between restriction sites also changes the size of restriction fragments. It is possible to detect some genetic diseases by analyzing the sizes of the DNA fragments produced by the cutting enzymes.

INTRODUCTION

Since the gross aspects of information organization in DNA are easily described by means of analogies between DNA and motion picture film, film metaphors are used now to begin describing gene manipulation. Imagine a film editor who wants to practice the fashionable technique of combining live and animated footage by clipping out a scene from a John Wayne movie and sticking it into a short Mickey Mouse cartoon. The spliced film will be longer than the original Mickey Mouse cartoon and, depending on which John Wayne scene was moved and where it was placed, the resulting movie will make more or less sense. The result will be the creation of a new motion picture. In the same sense, gene "splicing" creates new genetic arrangements.

Like the film editor, a gene cloner needs two tools: scissors and splicing tape. The scissors biologists use are specific enzymes that obey an important rule: they cut only at specific places in the DNA. The specificity arises because the cutting enzymes recognize certain short sequences of nucleotides in the DNA. If this rule were applied to motion picture editing, cuts could be made only where specific events occur in the movie. For example, the editor might have to follow a rule according to which the film may be cut ONLY where someone laughs, with cutting required anytime someone laughs in

ANY movie. Notice that the specificity rule leaves the editor simply holding the scissors, wielding them mindlessly wherever a given sequence of events appears. If nobody ever laughs in a particular movie, the editor would not be allowed to cut that film; consequently no splicing steps could be carried out. Obviously, however, people may laugh frequently in a given movie. By the rule, such a film must be cut into many pieces. In that case, an editor would have to sort through many John Wayne film fragments to find a particular scene for splicing.

As an alternative strategy, the editor could splice each John Wayne fragment into a separate copy of the Mickey Mouse cartoon, creating a large number of short, new movies, all of which would have to be viewed to find the desired scene. With film, the latter method is inefficient: technicians would have to make many splices that would be used only once; then editors would have to spend time screening all the new films to find the desirable one. Gene cloners, however, use this second strategy because they can easily create and examine thousands of "splicing" events and find a particular arrangement of nucleotides.

EcoRI   5' G A A T T C 3'
          3' C T T A A G 5'

RESTRICTION ENDONUCLEASES

Restriction endonucleases are a group of enzymes that correspond to the scissors in the analogies developed above. These **nucleases** occur naturally in a large number of different bacterial species, serving as part of the natural defense mechanism that protects bacterial cells against invasion by foreign DNA molecules such as those contained in viruses. Crucial to this protective device is the ability of the nuclease to discriminate between its own DNA and the invading DNA; otherwise the cell would destroy its own DNA. The recognition process involves two elements. First there are specific nucleotide sequences that act as targets for the nuclease. These are called **restriction sites.** Second, there is a protective chemical signal that can be placed by the cell on all the target sequences that happen to occur in its own DNA. The signal modifies the DNA and prevents the nuclease from cutting. Invading DNAs, lacking the protective signal, would be chopped by the nuclease. Restriction endonucleases from different bacteria often recognize different target sequences. Thus restriction endonucleases purified from different bacteria can be used as enzymatic tools to cut DNA at different, specific sites.

The different nuclease types recognize nucleotide sequences in DNA that are often four or six base pairs long. The enzymes then cut both strands of DNA. Some enzyme types cut within the recognition site while other types cut close to it. In some cases the two DNA strands are not cut opposite each other—rather, the cuts are staggered. In the case illustrated in Figure 7-1, the cuts are offset by four nucleotides. Once the cuts in this example have been made, only four base pairs remain between cut sites. Under most conditions, four base pairs are not enough to hold DNA together, and the DNA molecule separates into fragments (Figure 7-1).

LIGATION

As pointed out above, some restriction endonucleases generate staggered cuts (Figure 7-1b). The four nucleotides in the single-stranded ends of the DNA molecules are complementary to the ends of other molecules generated by cutting with the same restriction endonuclease. Thus, when two DNA molecules having complementary ends collide, the single-stranded ends temporarily form base pairs, and under suitable conditions the two molecules will temporarily stick together (Figure 7-2). The production of staggered cuts generates what molecular biologists call **sticky ends.** Chapter 3 mentioned DNA ligase, an enzyme that performs the essential function of joining DNA molecules after DNA replication (step 2 in Figure 3-4). If this enzyme is present when two DNA molecules having sticky ends happen to come together, it will repair the breaks that had been introduced by the restriction endonuclease. Thus, DNA joining can be accomplished by simply mixing together DNA molecules having complementary sticky ends and adding DNA ligase plus adenosine triphosphate **(ATP)** as a source of energy. Under the proper conditions, blunt ends can be ligated, a feature useful for joining ends created by restriction endonucleases that do not produce sticky ends.

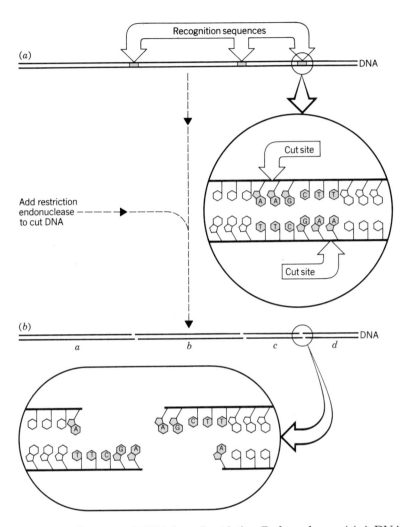

Figure 7-1 Cleavage of DNA by a Restriction Endonuclease. (a) A DNA molecule, depicted as two parallel lines, contains many short nucleotide sequences recognized by restriction endonucleases. **(b)** When a restriction endonuclease is added to the DNA, it binds to the DNA and cuts it. Some of these enzymes produce staggered cuts. The DNA molecule in the example is converted into four shorter molecules, *a, b, c,* and *d,* each with "sticky ends" that can form base pairs with each other. For clarity, the DNA molecules are not shown as interwound helices, and the hydrogen bonds between complementary base pairs are omitted.

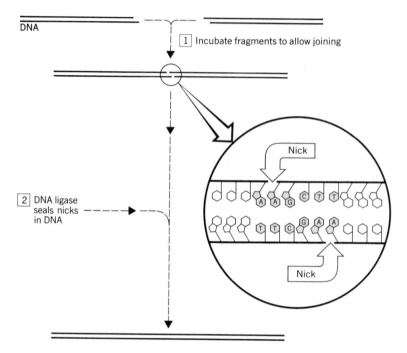

Figure 7-2 Joining Two DNA Fragments. (1) Two DNA molecules with complementary sticky ends are mixed and incubated. The molecules collide, and base pairs form. **(2)** The strand interruptions (nicks) are enzymatically sealed by DNA ligase.

CUTTING AND JOINING

The motion picture metaphor is useful when considering the details of cutting and joining DNA. First, imagine a large vat (representing a test tube) into which you have placed many, many unrolled copies of a particular John Wayne movie and a particular Mickey Mouse cartoon. The John Wayne film might represent many copies of an animal DNA, and the Mickey Mouse cartoon would represent many copies of a small DNA, the cloning vehicle, that can infect bacteria. Both are in solution in the same large receptacle. Now imagine that you and some friends enter the vat with splicing scissors and cut EACH film wherever anyone is shown laughing. This step corresponds to the addition to the test tube of many identical restriction endonuclease molecules to cut the DNAs. Soon, many fragments fill the vat. You and your friends leave the vat, so no more cutting can occur. The mixture is then stirred and the fragments collide. In the case of DNA in a test tube, some of the DNA collisions result in the ends sticking together, at least transiently. If DNA ligase is present, the fragments will become permanently joined.

As mentioned at the beginning of the chapter, the goal of gene "splicing" is analogous to obtaining one specific scene from the chosen John Wayne movie inserted into a COMPLETE Mickey Mouse cartoon. This chore can be simplified in two ways. First, before any cutting is done, tape the ends of the cartoon together to form a circle. Second, choose a cartoon in which someone laughs only once; then there will be only one cut in the cartoon. By starting with a circular cartoon that can be cut just once, you ensure that no cartoon fragments will be created; hence there will be no cartoon reattachment chores to do before completion of the new film. Eventually a John Wayne fragment will collide with one end of the cartoon, and the two can be joined. The other end of the cartoon will at some time collide with the free end of the John Wayne fragment already connected to one end of the cartoon. When this second joining event occurs, a circle will have been formed, containing a complete cartoon plus a John Wayne fragment. If the two taped-together ends are released, the film can be projected from beginning to end. The new film will make sense except for the brief interruption imposed by the John Wayne sequence, which probably would not confuse viewers of the cartoon any more than would a television commercial. A general scheme for DNA is outlined in Figure 7-3; in the case shown, the cloning vehicle occurs naturally as a circular plasmid.

In the vat, collisions between ends occur randomly, and much of the time the two ends of the cartoon simply rejoin. Likewise, John Wayne fragments attach to each other. In general, quite a mess is created. Occasionally, however, a single John Wayne

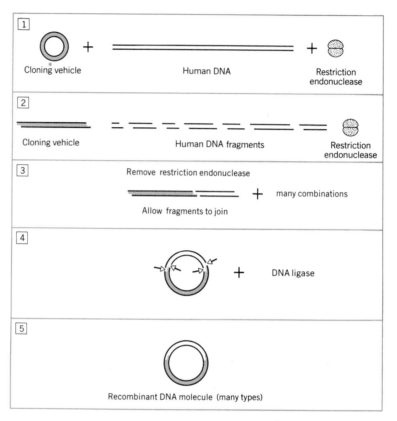

Figure 7-3 General Scheme for Forming Recombinant DNA Molecules.
(1) Circular plasmid DNA (cloning vehicle), human DNA, and a restriction
endonuclease are mixed. **(2)** Both DNAs are cut, producing a linear
plasmid and many human DNA fragments. In this example, all DNAs
in the mixture have complementary, sticky ends. **(3)** Occasionally a human
DNA fragment will attach to one end of the plasmid. Many combinations
form because many different types of human DNA fragment can join to
the plasmid. **(4)** Eventually both ends of the human DNA fragment will
have attached to the respective, corresponding ends of the plasmid. When
DNA ligase is added, the discontinuities in the DNA strands (arrows)
will be sealed, producing a circular recombinant DNA molecule with no
breaks in the DNA strands **(5)**. The ligation of different human DNA
fragments to plasmids causes the formation of recombinant molecules of
many types.

fragment will collide with the cartoon. But since there are thousands of different John Wayne fragments, only rarely will any PARTICULAR fragment attach to the cartoon. Thus, in genetic engineering, trillions of DNA molecules must be incubated together before there is a reasonable chance that the desired fragment will attach to the cloning vehicle.

RESTRICTION MAPS

In addition to their role in DNA cloning procedures, restriction endonucleases play an important part in the analysis of nucleotide sequences of DNA molecules (described in Chapter 10). The initial step in these analyses is called **restriction mapping;** the following procedure will provide familiarity with the cutting enzymes employed. First, return to the analogy between DNA and motion picture film, and imagine that you wish to study a film loop 1000 feet long. You are not allowed to use a movie projector, and the film has no ends: it is a circle. You could begin to examine this tangled mass if you could cut the film at SPECIFIC places to produce DISCRETE, manageable pieces. Now suppose that you have three kinds of scissors: type A, which cuts when somebody laughs; type B, which cuts when a dog bites a man; and type C, which cuts whenever a car door is opened (these scissors correspond to different restriction endonucleases cutting at specific sites on DNA). Assume that in the film you are studying somebody laughs only once. Therefore type A scissors will cut only once, producing two ends and giving you a reference point:

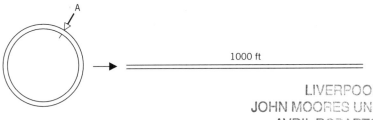

If in your film men are bitten by dogs three times, the circle will be cut into three fragments by type B scissors. You can measure the lengths. Suppose that the fragments are 100, 300, and 600 feet long.

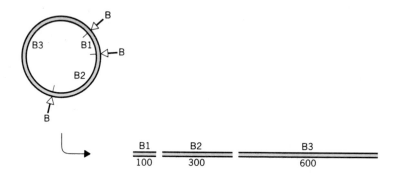

Likewise, if a car door opens twice in the film, type C scissors will cut two times, producing two fragments. Suppose that these fragments are 200 and 800 feet long.

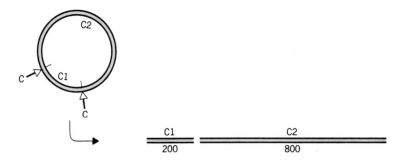

Even smaller fragments can be obtained by cutting with more than one scissors type. For example, type A and type B combined should produce four fragments. Suppose that you find the resulting lengths to be 50, 100, 300, and 550 feet. In such a case the type A cut must have occurred within the 600-foot fragment created by the type B enzyme (B3), producing two new fragments (50 and 550). There are only two ways the film can be arranged to produce this result, and in both cases the laugh (A) occurs between the two most widely spaced dog bites (B). Thus the map is beginning to develop. At this point you cannot tell whether cut A is nearer to fragment B1 or to fragment B2.

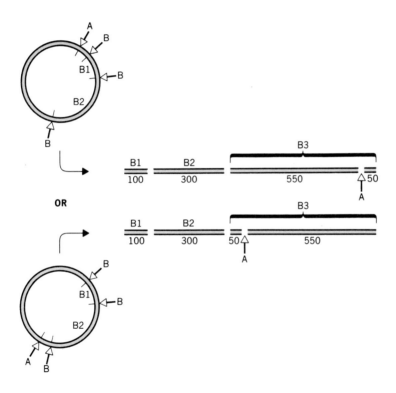

The next combination involves cutting with type A and type C scissors. This time you find three fragments, which are 200, 375, and 425 feet long. The type A cut must have occurred within the 800-foot C2 fragment, generating two new pieces 375 and 425 feet long.

Again there are two places where the A cut could be relative to the two C cuts.

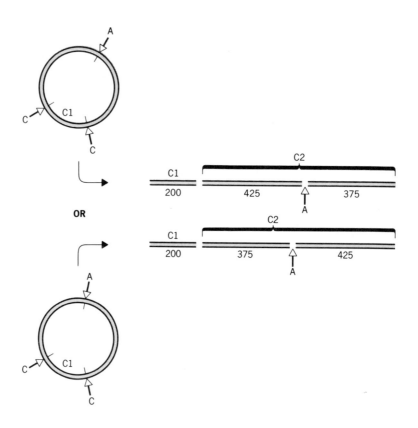

You can also cut with a combination of type B and type C enzymes. Suppose this produces five pieces, 75, 100, 125, 225, and 475 feet long. The 100-foot piece is B1, but B2 (300 feet) and B3 (600 feet) have disappeared. This must mean that one C-type cut occurs in B2 and the other in B3. The 75- and 225-foot fragments add up to 300 feet, which corresponds to B2; thus one C-type cut is 75 feet from one end of B2. The sum of 125 and 475 is 600, the value of B3; consequently, the other C-type cut is 125 feet from one end of B3. Since the two C-type cuts are either 200 or 800 feet apart on the accompanying circular map, there is only one way the map can fit together.

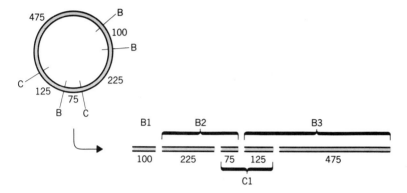

By adding the results from the A–C, A–B, and B–C combinations, it is possible to determine the position of the A cut. Since A is in B3, only 50 feet from a B cut, and far from the C cuts, A must map as follows:

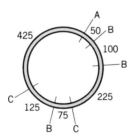

When all three types of cut are made simultaneously, six fragments will be produced.

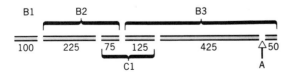

At this stage in the film analogy one would say that the sequence of events, beginning at A, is laugh, two dog bites, a car door opening, another dog bite, and finally another car door opening. The film distance between the events is also known.

In the case of DNA, the size of each of the fragments is measured by a technique called **gel electrophoresis**. A semisolid material similar to agar is first poured and allowed to solidify in the form of a slab. The teeth of a broad-toothed comb are inserted into the gel while still liquid. After solidification, the comb is removed, leaving a series of small wells near one end of the gel slab. A mixture of DNA fragments, produced by restriction endonuclease-induced cleavage, is then placed in a well. An electric field is applied across the gel, forcing the negatively charged DNA molecules to move through the gel. DNA molecules having the same size move together as a group. If trillions of identical DNA fragments are present, you can see them as a band by staining the gel. Smaller DNAs move faster through the gel than larger ones; consequently, fragments of different lengths produce bands at different positions in the gel (Figure 7-4). The size of DNA molecules in each band is determined by comparing how far a given band moves into the gel relative to bands of DNA molecules whose lengths are already known.

DNA in a band can be removed from the gel, inserted into a cloning vehicle, transferred into a bacterium, and amplified by growth of the microorganism. The nucleotide sequence of the DNA can then be determined as described in Chapter 10. By knowing the map order of the fragments (using the type of logic just outlined), it is possible to fit together the short nucleotide sequences from all the fragments like a jigsaw puzzle, eventually obtaining the nucleotide sequence of the entire DNA molecule.

Restriction mapping is also used as a diagnostic tool for prenatal detection of certain genetic diseases. For example, a single nucleotide change is responsible for sickle-cell disease (see Figure 3-5). This change, which occurs in a restriction endonuclease recognition site, renders the site unrecognizable by the enzyme; the enzyme does not cleave at that location. Consequently, the restriction map of DNA from a sickle-cell fetus differs from that obtained from normal DNA. The principle is illustrated in Figure 7-5 using the restriction mapping example given earlier in this chapter. In this example, the base change causing the disease occurs in the

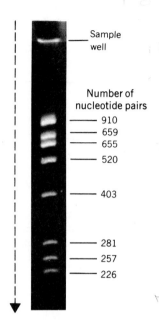

Sample well

Number of nucleotide pairs
—— 910
—— 659
—— 655
—— 520

—— 403

—— 281
—— 257
—— 226

Figure 7-4 Separation of DNA Restriction Fragments by Gel Electrophoresis. A mixture of DNA fragments was placed in a sample well (slot) in a gel and driven into the gel by an electric field. Smaller fragments move faster than larger ones, so when the electric field was turned off, the fragments had separated. The gel was stained to reveal the DNA bands and then photographed. Each band represents many identical DNA molecules having the same length. The direction of DNA movement is from top to bottom; the number of nucleotide pairs in each fragment is indicated. (Photograph courtesy of Richard Archer, Lasse Lindahl, and Janice Zengel, University of Rochester.)

recognition site for the type A restriction endonuclease, eliminating that site from the DNA. The map of the DNA from the diseased fetus would then be easily distinguishable from that of a normal fetus: two normal fragments (425 and 50) would be replaced by a new fragment (475) in the mutant DNA. Restriction fragment changes are usually detected by a method called Southern transfer hybridization (Chapter 8).

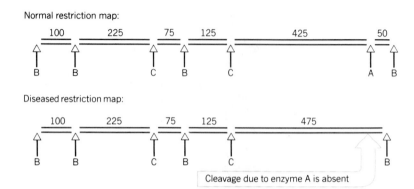

Figure 7-5 Diagnosing a Genetic Disease by Restriction Mapping. In this hypothetical example, a change in the nucleotide sequence in DNA causes a disease and also eliminates a known restriction site (the site for enzyme A). Thus DNA from a diseased cell has an altered restriction map. Cells from higher organisms such as humans have two copies of each DNA molecule, one derived from the mother and one from the father. Although these molecules are never identical over their whole length, they may be identical in the region being examined. If in this region DNA molecules derived from the parents are both normal or both pathogenic, the analysis produces the result shown in the corresponding (Normal/Diseased) diagrams. If, however, the DNA from one parent is normal but the DNA from the other parent is pathogenic, a mixture of the restriction fragments is observed. Sickle-cell disease is known to be due to a single nucleotide change in DNA (see Figure 3-5), and it is possible to diagnose the disease by examining restriction maps of fetal DNA.

There are many regions in human DNA that vary from one person to another such that the number of nucleotides (the length of the DNA) between two particular spots cut by a restriction endonuclease differs from person to person. This is sometimes due to short sequences being repeated different numbers of times, sometimes to sequences being added or subtracted from the region, and sometimes to a restriction site being created or eliminated in the region. Such regions that have different forms are called **restriction fragment length polymorphisms** or **RFLPs** (polymorphism means multiple forms). These regions do not necessarily occur in genes, and the function of most of them in the human **genome** is unknown. By carefully examining the DNA of members of families that carry genetic diseases, it has been possible to find forms of particular RFLPs that tend to be inherited with particular diseases. Generally the fragment differences occur not because a restriction site was created or disrupted by the diseased state itself (sickle-cell disease is an exception), but rather because the nucleotide sequence differences just happen to be near the gene involved. A particular form of a polymorphism that is close to a diseased gene tends to stay with that gene during the chromosomal rearranging and sorting phase that precedes fertilization and formation of the fetus. Thus RFLPs serve as markers of disease, and they have been useful for detecting maladies such as Huntington disease, cystic fibrosis, sickle-cell disease hemophilia, and certain types of colon and lung cancer.

Restriction analyses are also used to track down the source of infectious diseases. Hospital infections caused by the pathogenic bacterium *Staphylococcus aureus* represent a good example. Outbreaks of "hospital staph" often occur in nurseries and cardiac units, and because these strains are frequently resistant to many antibiotics, they have become a very serious problem. Locating and eliminating reservoirs of the infectious organism is a major challenge. In one case, restriction fragment analyses of bacteria taken from surgical patients suffering from postoperative toxic shock syndrome, one of the consequences of infection by *S. aureus,* showed beyond doubt that a surgeon had been harboring the bacteria in his nose. Once this became known, steps were taken to rid the surgeon of his symptomless infection.

VISUALIZATION OF CLONED DNA

Most of the basic tools used in gene cloning have now been introduced: the restriction endonucleases that cut DNA molecules in specific places, the ligases that join DNA molecules, and the plasmids and phages that carry new combinations of DNA molecules into cells where they can be reproduced (the nucleic acid probes used to identify specific colonies or plaques that contain cloned genes are

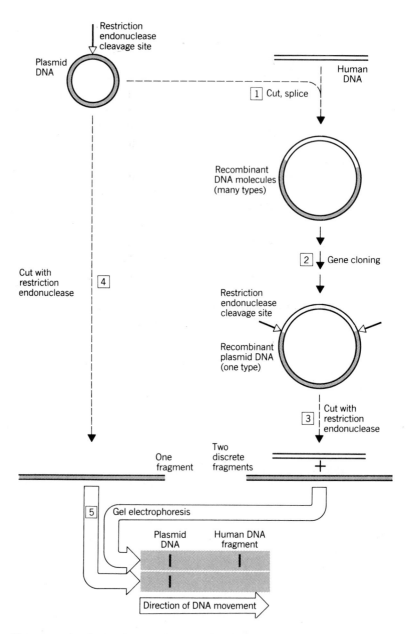

Figure 7-6 Analysis of Recombinant DNA by Gel Electrophoresis. (1) Plasmid and human DNA molecules are cut and joined to form recombinant plasmid DNA molecules. Many combinations of DNA fragments join, producing many different recombinant DNAs. **(2)** The recombinant DNAs are

described in the next chapter). At this point it is useful to consider physical evidence that new DNA has actually been inserted into a phage or plasmid. One type of proof for insertion involves changes in restriction maps. Figure 7-6 illustrates how a human DNA fragment changes the restriction map of the plasmid into which it has been cloned. In this example, the plasmid contains a single site where a particular restriction endonuclease cuts. At this site human DNA fragments, generated by the same type of endonuclease, are inserted into the plasmid molecules to form recombinant DNA molecules (step 1, Figure 7-6). The recombinant DNA molecules are introduced into bacterial cells, individual clones are obtained (see Chapter 9), and recombinant DNA molecules are purified. This produces a single type of recombinant plasmid (step 2). When these DNAs are cut with the restriction endonuclease (step 3), two pieces of DNA are created. These two pieces can be analyzed and compared with the single piece generated by cleaving the original plasmid (step 4), using gel electrophoresis (step 5).

PERSPECTIVE

DNA molecules are so long and contain so much information that until the discovery of restriction endonucleases there seemed to be little hope of determining extensive nucleotide sequences. Now that DNA molecules can be cut into discrete, manageable fragments,

---

introduced into bacterial cells, and individual colonies are grown. A bacterial colony containing cloned DNA is identified (Chapter 9), and recombinant plasmid DNA is isolated from it. (All the recombinant DNA molecules from a colony are identical. They all have two restriction endonuclease cleavage sites, one at each junction between human and plasmid DNA.) **(3)** Cleavage of the recombinant plasmid DNA with restriction endonuclease produces two discrete DNA fragments. **(4)** The original plasmid contains only one restriction endonuclease cleavage site, so it is cleaved only once by the nuclease. **(5)** The products of steps 3 and 4 are analyzed by gel electrophoresis, as described in Figure 7-4. **(5)** The original plasmid produces only one band; the recombinant plasmid produces two.

determining nucleotide sequences has become routine. Attention is currently focused on determining the complete nucleotide sequence of the human genome, an effort comparable in scope, scale, and funding to that required to put a man on the moon. The next chore will be to discover what the various sequences do. Some information can be obtained by examining how specific, cloned sequences function when placed inside living cells, especially if the sequences encode proteins. But the task becomes massive when one begins to ask detailed questions about how different regions of DNA interact to coordinate the control of gene expression.

Restriction fragment analyses have many practical applications. Already in place are tests to identify fetuses destined to suffer from genetic diseases and adults who are predisposed to certain types of heart disease and cancer. Such analyses can also be used in general identification processes, and law enforcement agencies are taking advantage of the technology. More controversial issues, just now coming into the political arena, include screening for insurance coverage and employment suitability. Some of these issues are discussed in the last chapter of this book.

Questions for Discussion

1. The molecular scissors used to cut DNA are called restriction endonucleases. What does endonuclease mean? (Define the word fragments *endo-*, *nucle-*, and *-ase*.)
2. Restriction endonucleases are extremely specific in terms of their nucleotide recognition sequence. How does their natural function in cells help explain this high degree of specificity?
3. Suppose that you are studying a new plasmid that is 2500 base pairs long, and you wish to construct a restriction map. You treat the plasmid DNA with a set of restriction endonucleases and measure the size of the resulting DNA fragments by gel electrophoresis. You obtain the following results (the abbreviations on the left refer to specific restriction endonucleases; the

numbers are the sizes of the DNA fragments in nucleotide pairs):

*Eco*RI—2500
*Hind*III—2500
*Pst*I—2500
*Mbo*I—1300, 800, 400
*Mbo*I–*Eco*RI—1300, 600, 400, 200
*Mbo*I–*Hind*III—1300, 800, 300, 100
*Mbo*I–*Pst*I—1000, 800, 400, 300
*Eco*RI–*Hind*III—2000, 500
*Eco*RI–*Pst*I—1600, 900
*Hind*III–*Pst*I—2100, 400

Construct a map based on the information given above. In your map place base pair 1 at the *Hind*III site.

4. Suppose that you had two restriction endonucleases, *Eco*RI and *Mbo*I, each in a separate test tube. Somehow the labels came off the tubes. How could you determine which tube contained *Eco*RI using the plasmid described in question 3?

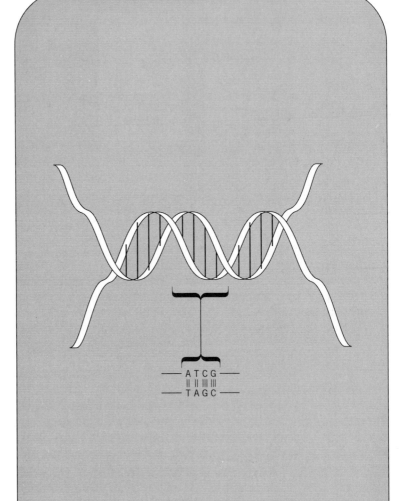

Complementary base pairing is the key to nucleic acid recognition.

USING COMPLEMENTARY BASE PAIRING

Hybridization, Probes, and Amplification

Overview ─────────────────────────────────

Attractive forces between complementary base pairs in DNA are disrupted when DNA is heated, but the base pairs re-form if the DNA is cooled slightly and incubated long enough for the complementary strands to find each other. Single DNA strands from different sources will form hybrid duplexes if mixed and incubated appropriately. Hybrid formation is a measure of relatedness, and it can be used to search large populations of bacterial cells or phage plaques for specific recombinant DNA molecules. Since hybridization can reveal the presence of particular nucleotide sequences in a DNA molecule, it has been important for tracking DNA from one generation of a family to the next. The principle of complementary base pairing and an understanding of DNA polymerase have also led to a test tube method for amplifying short regions of DNA many, many times. Complementary base pairing is the heart of DNA technology, just as it is central to DNA biology.

INTRODUCTION

Most of the tools used in gene cloning have been presented. The cloning vehicles that carry DNA fragments into cells are themselves specialized DNA molecules (plasmids and phages) that can replicate inside microorganisms; restriction endonucleases and ligases make possible the insertion of DNA fragments into the cloning vehicles; and the ability of tiny microorganisms to form colonies makes it possible to separate and fish out desired recombinant DNAs. The remaining step to discuss is the identification of the colonies or plaques harboring a particular recombinant DNA.

The key to identifying **nucleic acids** rests on the principle of complementary base pairing. As pointed out in earlier chapters, the process can be represented by a simple rule: when two nucleic acids form a duplex, G pairs with C and A pairs with T or U. Only when two nucleic acids can become extensively base-paired will they fit together and form a double-stranded molecule. If one nucleic acid, called the **probe,** is marked in an easily detectable way, such as with radioactive atoms, it can be used to find another nucleic acid: when the two form a duplex, the signal given off by the probe tells the investigator where the target molecule is and how much is present. This general process of forming double-stranded nucleic acids is called **nucleic acid hybridization.**

Hybridization is the key to locating cloned genes within a large population of bacterial cells or phages. It is also used extensively to study the biology of nucleic acids. A third application involves amplification of DNA fragments by a revolutionary process called the **polymerase chain reaction (PCR).** These three aspects of complementary base pairing are described after a brief definition of hybridization.

NUCLEIC ACID HYBRIDIZATION

When double-stranded DNA is dissolved in dilute salt water and gradually heated, the two strands separate at a temperature slightly below boiling. Strand separation, which is often called denaturation or melting, occurs over a range of about 5°C. If the temperature is lowered to about 25°C below the melting temperature, the complementary DNA strands will reassociate. The general idea is that two strands will occasionally bump into each other and form a short region of double-strandedness. Then nearby nucleotides pair, and the strands rapidly zip together.

If DNA molecules from two related organisms, such as monkeys and humans, are mixed, heated, and then cooled, occasionally a DNA strand from one primate will hybridize with a DNA strand from the other. Since the nucleotide sequences of human and monkey DNA are not identical, the hybrid duplex formed will not be perfectly paired. Nevertheless, the pairing will be good enough to hold the strands together. If the temperature is then raised to melt the monkey : human hybrid, the imperfect pairing will cause the strands of the hybrid to come apart at a slightly lower temperature than observed for a monkey : monkey or a human : human duplex.

If RNA is mixed with DNA and the solution is heated to melt the DNA, cooling will allow the RNA strands to form hybrids with regions of the DNA that are complementary to the RNA. Two RNA strands will even form hybrids if they are complementary.

One way to simplify measurement of relatedness is to melt the DNA and then attach it to a paper (nitrocellulose) filter while the DNA is still single-stranded. Then the filter is placed in a mixture of **radioactive,** single-stranded DNA (or RNA) at roughly 25°C below the melting temperature. If the radioactive DNA, the probe, is complementary to the DNA attached to the filter, it will form double-stranded molecules with it (Figure 8-1). Those molecules will remain double-stranded when the filter is washed to remove any unreacted radioactive DNA. The amount of radioactivity associated with the filter after washing will reflect the number of hybrids formed.

USING PROBES TO FIND CLONED GENES

As pointed out in the introduction to this chapter, cloning vehicles are used to carry a collection of DNA fragments into bacteria or yeast, which are spread onto an agar plate to physically separate the DNA fragments. The cells grow into colonies, and the ones containing the fragment of interest must then be identified. Colony hybridization is carried out in the following way. First, a piece of filter paper is placed on the agar plate long enough for some of the cells in each colony to stick to the paper. Then the paper is lifted

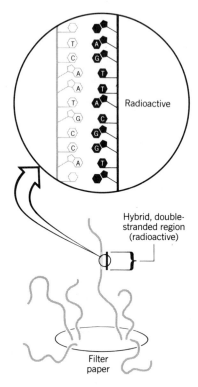

Figure 8-1 Nucleic Acid Hybridization. Under the appropriate conditions
two complementary single-stranded nucleic acids will spontaneously form
base pairs and become double-stranded. If single-stranded, nonradioactive
DNA (shaded) is fixed tightly to a filter and then incubated in a solution
containing single-stranded, radioactive DNA (solid), double-stranded regions will form in which the two types of DNA have complementary nucleotide sequences; the radioactive DNA will become indirectly bound to the
filter through its attachment to a specific region of nonradioactive DNA
(open). By measuring the amount of radioactivity bound to the filter, one
can estimate the relatedness of two DNAs. Complementary base pairing
between RNA and DNA produces an **RNA:DNA hybrid.**

146

off, and some of the cells are also removed. The specific pattern of colonies on the plate is preserved on the paper. Next, the cells on the paper are broken in such a way that the DNA sticks to the paper. During this process the double-stranded DNA is converted into single-stranded molecules, thus making the bases available for base pairing. Then the paper, with its attached spots of cellular DNA, is placed in a solution containing single-stranded, radioactive probe DNA (or RNA) known to be complementary to the gene to be isolated. The probe forms base pairs only with DNA from the particular colony that contains the gene being sought. The cellular DNA becomes indirectly radioactive when hybridized to the radioactive probe. Clusters of broken, radioactive cells on the filter paper can be identified by placing the paper next to X-ray film, for the radioactive emissions will expose the film. Since the patches of cells on the paper will be in the same pattern as the colonies on the agar plate, the location of the radioactivity on the paper indicates which colonies on the agar plate contain the cloned gene. Those colonies can be picked with a wire loop and grown on fresh agar plates.

The key to finding the right colony is having the correct probe. Thus it is important to consider how the probes are obtained. One way involves purifying the protein product of the gene being sought, followed by chemical analysis to determine the order of the amino acids in the protein. Since every amino acid corresponds to a triplet of nucleotides in the DNA of the gene, it is possible to predict the nucleotide sequence of the gene from the sequence of amino acids (the prediction is not exact because most amino acids are encoded by several different codons; see Figure 2-6). The next step is to chemically synthesize a short stretch of DNA, one base at a time, that will have a sequence identical to that predicted for a region of the gene. If the probe is long enough (20–30 nucleotides), it will hybridize to the gene even though the probe may not match the gene perfectly (most of the amino acids are encoded by more than one codon, making it difficult to obtain the exact nucleotide sequence from the amino acid sequence). An alternative strategy is to make a mixture of all the oligonucleotides that could possibly encode the region of the protein for which the amino acid sequence is known. One of the species in the mixture would match the gene exactly. In either case, a purified enzyme called a **kinase** is used to add radioactive phosphorus to the synthetic DNA fragment, making it a highly radioactive probe.

In rare cases it is also possible to isolate messenger RNA for the gene that is to be cloned. In principle, cells grown in the presence of radioactive nucleotides will yield radioactive messenger RNA suitable for use as a probe. However, it is often difficult to obtain natural messenger RNA containing enough radioactivity to test the bacterial colonies for cloned genes—so much radioactivity must be added to the cells that they tend to die. Consequently, the messenger RNA is usually employed as a template for the synthesis of DNA using **reverse transcriptase,** an enzyme purified from **RNA tumor viruses.** Since the DNA product, called **complementary DNA** (cDNA), is synthesized in test tubes, it can be made highly radioactive by using radioactive nucleotides to form the DNA.

Once a small fragment has been cloned, it can be used as a probe to study the large DNA molecule it came from. Examples are given in the next section. It can also be used to clone a fragment of DNA adjacent to its position in the long DNA. This process is called walking along DNA and is described later in this chapter.

TRANSFER HYBRIDIZATION

Very small alterations in the nucleotide sequence of DNA molecules can be detected by analysis of changes in the location of cuts by restriction endonucleases. These changes alter the size of the DNA fragments produced and thus the restriction maps of the DNA molecules (see Figure 7-5 for a hypothetical example). Since large DNA molecules have very complicated restriction maps, a method called transfer hybridization has been devised to examine only a small, specific region of the entire DNA. In this method, a radioactive probe, obtained by cloning, is used in the identification, by means of nucleic acid hybridization, of specific DNA fragments. The procedure is outlined in Figure 8-2. After DNA has been cut by a restriction endonuclease and the fragments separated from one another by gel electrophoresis, the fragments are **denatured** and transferred from the gel to nitrocellulose paper. There they can be hybridized to a radioactive probe that contains nucleotide sequences from the gene or region of interest. After incubation with the probe, nonhybridized radioactivity is washed off the paper, and the paper is placed next

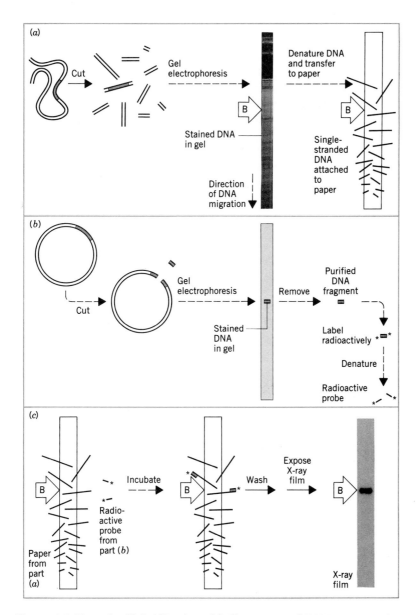

Figure 8-2 Transfer Hybridization. (a) Chromosomal DNA is cut with a restriction endonuclease to produce small DNA fragments. The gene of interest is shown as a solid region, and in this example it is found on only one type of fragment. The fragments are separated by gel electrophoresis.

(Continued)

149

to a piece of X-ray film to expose it. Since the relative position of each fragment is maintained during the transfer from gel to paper, the position of the radioactive band can be used to calculate the size of the fragment containing nucleotide sequences identical to those in the probe.

Transfer hybridization, combined with cleavage of DNA by restriction endonucleases, has become a popular tool for prenatal diagnosis of certain genetic defects. In Chapter 7 it was pointed out that genetic diseases are sometimes associated with changes in restriction maps. Transfer hybridization makes it possible to limit the analysis to the region at or near the gene responsible for the disease.

Transfer hybridization can also be used to detect changes in the location of specific regions of DNA. For example, the method was used to show that two strains of the same bacterial species contain *tst*, the gene for toxic shock syndrome, at different locations in their respective chromosomes (Figure 8-3). Could this mean that the *tst* gene is able to jump around in the DNA?

WALKING ALONG DNA

Molecular biologists often want to obtain adjacent regions of DNA from a piece that has been cloned. This can be accomplished by a modification of the cloning procedure called **walking along DNA.** The general strategy is to use a nucleotide sequence near one end of the cloned region as a probe for locating adjacent, overlapping

---

Figure 8-2 (*Continued*) Many bands (fragments) are seen following staining of DNA (band **B** contains the gene of interest). The DNA in the gel is denatured and transferred to nitrocellulose paper. **(b)** Cloned DNA containing the gene of interest is purified and cut with a restriction endonuclease to release a small fragment from within the gene of interest. This small fragment of DNA is the probe; it is purified by gel electrophoresis, radioactively labeled, and denatured. **(c)** The filter-bound DNA and the probe are mixed and incubated. After band **B** has hybridized to the probe and become radioactive, the nonhybridized radioactive DNA is removed. Band **B** is easily identified on the film. Its position on the paper is related to its length (smaller DNA will be nearer the bottom).

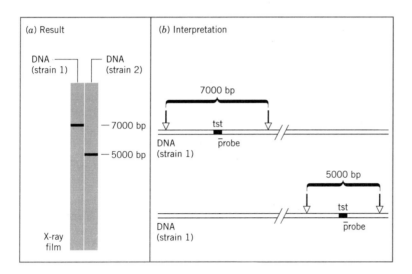

Figure 8-3 Location of Toxic Shock Syndrome Gene Varies Among Strains. DNA isolated from two strains of the bacterium *Staphylococcus aureus,* the organism that causes toxic shock syndrome, was examined as described in Figure 8-2. The DNA samples were digested with the restriction enzyme called *Cla*I, an enzyme that does not cut inside the toxic shock syndrome toxin-1 gene *tst*. The fragments were next subjected to **electrophoresis** in an agarose gel. The gel was treated with alkali to denature the DNA and then placed on nitrocellulose paper to transfer the DNA fragments by a blotting technique. Following the transfer, the paper was incubated with a radioactive probe obtained from the cloned *tst* gene. After rinsing, the paper was used to expose X-ray film, with the result shown in **(a)**: a single band of DNA containing 7000 base pairs (bp), for one DNA sample and a single band of 5000 bp for the other. To obtain this result, the DNA surrounding the *tst* gene had to be different in the two strains. The gene was probably at different positions in the chromosomes of the two strains: arrows in (b) indicate the locations of restriction endonuclease cleavage sites. Perhaps *tst* is associated with a transposon (see Chapter 11) that can move from one spot on the chromosome to another. (Photo courtesy of Barry Kreiswirth and Richard Novick, Public Health Research Institute.)

151

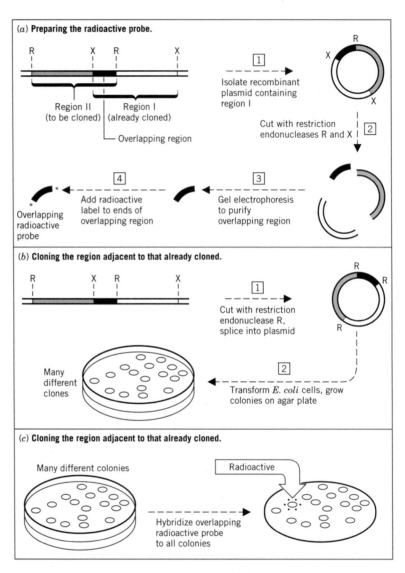

Figure 8-4 Walking Along DNA. (a) Preparing the radioactive probe. A short piece of DNA is shown with two adjacent, partially overlapping regions. Region I has already been cloned and is bounded by cleavage sites for restriction endonuclease X. Region II, the portion to be cloned is bounded by cleavage sites for restriction endonuclease R. **(1)** Region I DNA is inserted into a plasmid. **(2)** Region I–plasmid recombinant DNA is cut by endonucleases R and X to liberate the short piece of DNA (solid) that represents the

regions in a collection of recombinant DNAs (Figure 8-4). Two restriction endonucleases are used. First, one is found that cuts at or near an end of the cloned fragment (X in Figure 8-4a). Then another is found that cuts at a site in the cloned sequence and also at a site far outside the cloned region (R in Figure 8-4a). When human DNA is cut with enzyme R only and the fragments are cloned into plasmids, one type of recombinant DNA (region II, Figure 8-4a) will partially overlap with the original cloned sequence. The overlapping sequence is represented by the solid region in Figure 8-4. Bacterial colonies containing this DNA are identified by the colony hybridization technique using a radioactive probe from the overlapping region obtained as outlined in Figure 8-4. The DNA fragment (region II, Figure 8-4) can be isolated from recombinant plasmids by methods involving gel electrophoresis. By repeating this process, one can move along a DNA molecule, successively cloning small bits of DNA.

DNA AMPLIFICATION BY PCR

Several of the principles derived from studies of DNA replication (Chapter 3) and complementary base pairing have led to a method

---

overlap between regions I and II. (3) The overlapping DNA piece is separated from all other pieces by gel electrophoresis (see Figure 7-4). (4) Radioactive label is enzymatically attached to the overlapping fragment. (b) Cloning the region adjacent to that already cloned. (1) Total DNA is cut with restriction endonuclease R and is inserted into plasmid DNA, producing recombinant DNA molecules of many types. (Only the particular recombinant DNA being sought is illustrated. In this DNA the region of overlap is present because endonuclease X has not been used.) (2) *E. coli* cells are transformed with the recombinant plasmids, and these cells are grown into colonies on agar plates. Very few of these colonies contain the adjacent region (region II). (c) Identifying the colony containing the adjacent region. The many colonies obtained in (b) are tested by nucleic acid hybridization using the overlapping radioactive probe prepared in (a). The colony containing region II will become radioactive, and it can be identified by exposure of X-ray film. All plasmid DNA isolated from this colony will contain region II.

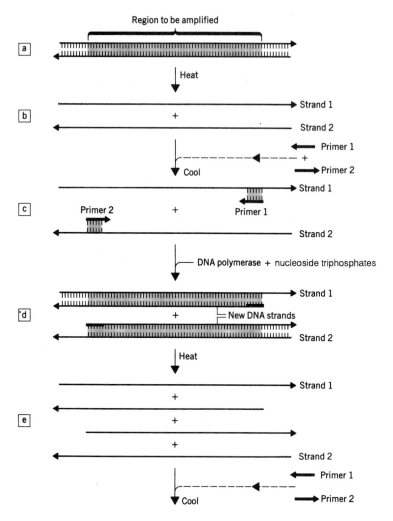

Figure 8-5 Amplifiction of DNA by PCR. (a) A double-stranded DNA molecule that contains a specific region of interest (double-strandedness is indicated by vertical lines). **(b)** Heating the DNA causes the two strands to separate. **(c)** When primers complementary to short regions on each of the two strands shown in (a) are added to the mixture and it is cooled, the primers hybridize to the two strands labeled **1** and **2**. **(d)** The primers are extended by the addition of DNA polymerase and nucleoside triphosphates. The new DNA molecules have different lengths because the polymerization

154

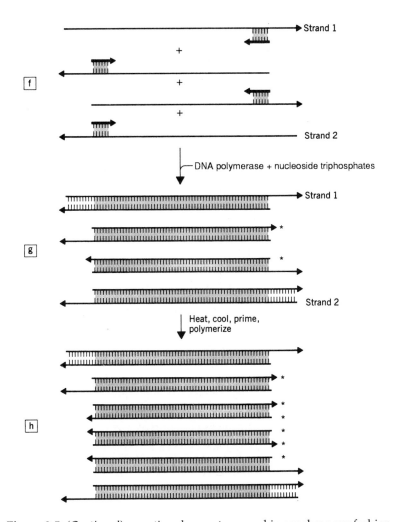

Figure 8-5. (*Continued*). reaction does not proceed in synchronous fashion. **(e)** If the mixture is heated, all the DNA will become single-stranded. **(f)** Upon cooling, primers will hybridize to both old and new strands in the mixture. **(g)** DNA polymerase again extends the primers. If the templates are the strands made in step d, DNA synthesis stops when the polymerase reaches the end of the template. This produces a discrete DNA fragment (*) containing the nucleotide sequences of both primers and the DNA in between. **(h)** A subsequent cycle of heating, cooling, and polymerization increases the relative abundance of the discrete fragment (*). Heat-resistant DNA polymerase is used. Therefore, if enough polymerase, nucleoside triphosphates, and primers are added at the beginning of the reaction, the process will consist of only heating and cooling steps.

155

for selectively synthesizing short regions of DNA. This procedure, the polymerase chain reaction (PCR) has had a revolutionary effect on DNA science because it can sometimes bypass the tedious, time-consuming steps required for standard cloning of DNA fragments. In the process, a specific segment of DNA can be enriched by more than a hundred thousand times relative to nearby nucleotide sequences. This process is outlined in Figure 8-5.

A key concept is that to begin DNA synthesis, DNA polymerase always requires a primer: that is, there must be available an end of a preexisting segment of either DNA or RNA to which DNA polymerase can add nucleotides. Thus DNA polymerase must start at points on a single-stranded DNA defined by primers. If the two strands of a long, double-stranded DNA molecule are separated and a short primer is hybridized to one strand at a specific spot (Figure 8-5c), DNA synthesis will always begin there. Because of the directionality of polymerase movement along DNA, synthesis can proceed in only one direction from the primer. A different primer can be placed on the second strand to permit the synthesis of the region between the primers (see Figure 8-5d). After DNA synthesis has occurred, the mixture is heated to produce single-stranded molecules from the newly made double-stranded ones. Upon cooling, the primers will hybridize to the new DNA as well as to the original strands. Another round of DNA synthesis generates discrete DNA fragments, which include the sequences of both primers and the DNA in between (Figure 8-5g). The cycle of heating, cooling, and polymerization, which takes only a few minutes, is repeated many times. With each cycle, the discrete fragment increases in abundance. Within 3 hours it is possible to obtain a specific fragment of DNA that can be easily cloned or purified by gel electrophoresis. The limiting factor is information about the nucleotide sequence: the biologist must know enough about the sequence to generate the primers.

One major application of gene amplification is in genetic screening. The amplification process generates a specific DNA fragment whose nucleotide sequence can be easily determined, making it possible to detect the presence of mutations associated with genetic diseases. The method can also be used to detect viral diseases such as AIDS. The nucleotide sequence of the viral genome is known, and if it is present in even one in a thousand human cells, it can be detected by amplification. A third application concerns the identification of people. Although all humans have very similar nucleotide sequences in their DNA, each person's DNA is unique. Thus amplification and nucleotide sequence determination of particularly variable regions of DNA can replace fingerprinting. The method is so sensitive that blood stains, skin fragments, and hair cells may provide enough material for analysis. Amplification will become increasingly important in criminal cases, for it is greatly increasing the ability of law enforcement agencies to demonstrate suspects' presence at crime scenes. It is also helping to establish innocence, particularly in cases of rape.

PERSPECTIVE

Since its first formulation in the early 1950s, complementary base pairing has assumed a central position in molecular biology. It is the process by which nucleic acids recognize each other, and it is the basis for copying genetic information. By the mid-1960s, complementary base pairing was being used to measure relatedness among organisms. A decade later gene cloning technologies emerged, and complementary base pairing became the basis for locating bacterial colonies or phage plaques harboring specific DNA fragments. By the late 1980s complementary base pairing was being used to amplify regions of DNA hundreds of thousands of times by the method called PCR. With each new application it became easier and easier to participate in DNA science. Now DNA analyses are routine forensic tools. Will the next step be gene testing kits for home use?

Questions for Discussion

1. When double-stranded DNA molecules are heated sufficiently, the two strands separate. Often strand separation occurs within a narrow temperature range. The temperature at which denaturation (melting) occurs depends on the strength of the forces

holding the strands together. If you form **hybrids** between two nucleic acids that are not perfectly complementary, the denaturation temperature will be lower than that of a perfect hybrid by one Celsius degree per 1% base pair mismatching. If hybrids formed between two related organisms have a denaturation temperature 10°C below that seen with duplex DNA of either organism and each organism has 1 million base pairs of DNA in its genome, how many bases are not paired in the hybrid?

2. In the examples given in the discussion of transfer hybridization, DNA was transferred from the gel to the filter. In such cases the transfer hybridization procedure is sometimes called **Southern blotting** in honor of E. Southern, the developer of the method. (The "blotting" part refers to the role of paper towels in the transfer process. Nitrocellulose paper was placed on top of the gel, and then a pile of paper towels was placed on top of the nitrocellulose. The towels "sucked" water through the gel and nitrocellulose, and along with the water came the DNA fragments. When they hit the nitrocellulose, they stuck tightly.) As you might expect, RNA can also be transferred from gels to filters, with the hybridization being carried out in a similar way. This type of experiment is called **Northern blotting** for reasons of literary parallelism. What can Northern blotting tell you about particular mRNA molecules extracted from cells?

3. PCR involves hybridization of primers, DNA synthesis by DNA polymerase, and heating to denature the primer–template DNA hybrids. During the heating step the temperature approaches that of boiling water, which will permanently inactivate most DNA polymerase molecules. Thus for each cycle you would have to add new DNA polymerase. The whole process is greatly simplified if the DNA polymerase survives the heat treatment. How would you find a heat-resistant DNA polymerase?

4. After amplification by PCR, the ends of the fragments often have short single-stranded tails that make it difficult to directly insert the fragment into a plasmid. The tails tend to be rich in As, but they are not necessarily complementary to sticky ends created by restriction endonucleases. How would you fix the ends of the fragments to facilitate cloning?

5. One practical application of gene cloning is detection of genetic abnormalities in fetuses. Develop a strategy to detect a disease that causes a change in the size of a specific restriction fragment using transfer hybridization. What materials (cloned genes, etc.) must be available to you? How could your system be used to detect carriers of the disease?

6. Suppose you know the nucleotide sequence of a particular gene in chimpanzees and wish to clone the **homologous** gene from human DNA. How would you use the polymerase chain reaction to help you clone the gene?

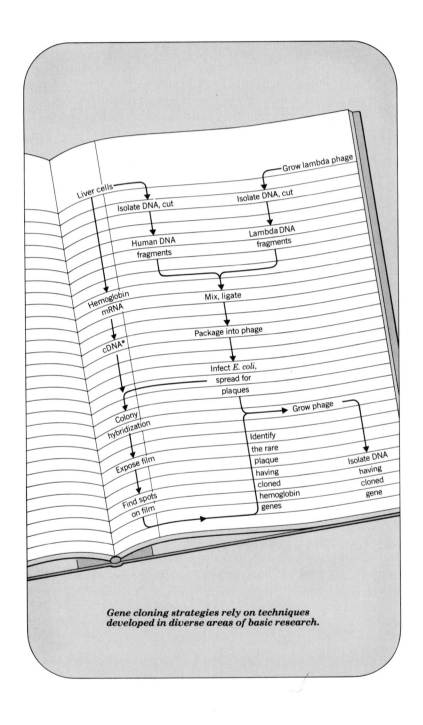

Gene cloning strategies rely on techniques developed in diverse areas of basic research.

CLONING A GENE
Isolation of a Hemoglobin Gene

Overview _____

There are now many strategies for cloning genes. In each case small fragments of DNA are inserted into cloning vehicles (i.e., into plasmid or viral DNA molecules). The vehicles carry the fragments into living cells, where the vehicles and their attached fragments reproduce. Cells containing plasmids grow into individual colonies, while those infected with a virus generate individual plaques. Both colonies and plaques can be tested for the presence of a particular piece of DNA by nucleic acid hybridization using a radioactive probe.

Gene cloning strategies with plasmids often use antibiotic resistance properties to help locate bacterial colonies into which genes have been cloned. One type of strategy employs plasmids having genes for resistance to two different antibiotics. One antibiotic resistance gene contains a restriction endonuclease cleavage site at which DNA fragments can be inserted; insertion of a fragment into this site on the plasmid destroys that resistance gene. The second antibiotic resistance gene is not affected. Thus recombinant plasmid DNA molecules will confer resistance to the second antibiotic only. When bacterial cells are transformed with plasmids thought to be recombinants, colonies are obtained that grow on agar containing the second antibiotic but not on agar containing the first antibiotic. Such colonies are then tested for the presence of a specific gene, often by nucleic acid hybridization.

Bacteriophage lambda DNA can generally accommodate longer inserts, so fewer recombinants must be screened to find one carrying the gene of interest. Analysis of the long insert then involves breaking it into smaller pieces that are recloned, a process called subcloning.

___

INTRODUCTION

As pointed out earlier, gene cloning is much like baking a cake; recipes are available for both processes. This chapter describes some general cloning strategies to tie together the concepts developed thus far. The focus is on the hemoglobin genes, the first mammalian genes to be cloned. Hemoglobin is the blood protein responsible for moving oxygen from the lungs to the body tissues, and a number of genetic disorders have been associated with defective hemoglobin. Among these is the serious hereditary disease called sickle-cell disease. The hemoglobin genes have also elicited interest because they code for separate proteins that function at different stages of our lives. Biologists want to understand how one hemoglobin gene is switched on and how another is switched off. Having the hemoglobin genes cloned makes many new experiments possible. An outline of the procedures is sketched in Figure 9-1.

Before discussing cloning methods, it is useful to consider the magnitude of the task. A human cell contains 6 billion base pairs of DNA (sperm or egg alone contains half that amount), which is experimentally broken into fragments about 20,000 base pairs long. Thus when recombinant DNA molecules are prepared by inserting DNA fragments into vector DNA, a collection (library) of at least 300,000 must be examined to find a particular recombinant. Several times this number are usually examined, however, since the recombinants are created randomly and there is a chance that a particular insert will not be included in the 300,000. Thus one would plan to examine a million clones.

OBTAINING DNA

The first step in gene cloning is to obtain DNA that contains the gene of interest. The initial hemoglobin studies were done with rabbits, but similar procedures would work for humans. With few exceptions, all the cells in a rabbit contain identical DNA molecules. Thus DNA from most rabbit cells can serve as a source for hemoglobin genes, and almost any type of body tissue can be ground up to obtain these genes. The original recipe called for extracting DNA from rabbit livers. Liver tissue was frozen, placed in a blender, and chopped until the cells were broken to release the molecules held inside. DNA was then purified by procedures similar to those outlined in Chapter 2. A detergent solution was added to unfold cellular proteins, and the cell lysate was treated for several hours with a protease, an enzyme that cuts proteins into fragments. Then phenol was added to the lysate. Although this oily substance and the watery cell lysate do not mix, vigorous shaking of the combination creates small droplets of phenol inside the water phase. Proteins that escaped enzymatic digestion moved into the phenol. When the mixture was subjected to centrifugation, the phenol and the cell lysate formed distinct layers in a test tube. The DNA-containing layer was removed and saved. By this stage most of the cellular proteins had been removed from the DNA preparation. The rabbit DNA was further purified by centrifugation in a density gradient as described in Chapter 6. The DNA formed only a single band in the centrifuge tube; it was easily removed with a syringe and placed in another test tube. Hemoglobin genes, millions of them, were in the test tube. But so were millions of copies of a hundred thousand other genes.

CLONING INTO BACTERIOPHAGE LAMBDA DNA

Hemoglobin genes were initially separated from most of the other genes in the DNA preparation by cloning them into bacteriophage lambda DNA (see Chapter 6 for description of bacteriophages). This phage is particularly attractive for cloning because large segments of foreign DNA can be inserted into its DNA without interfering with infectivity; thus the resulting clones might contain several adjacent genes as well as nearby noncoding regions. In this part of the recipe lambda phage particles were first purified by a method similar to that described in Figure 6-7. Next, DNA was extracted from the phage particles, using many of the same steps described above for

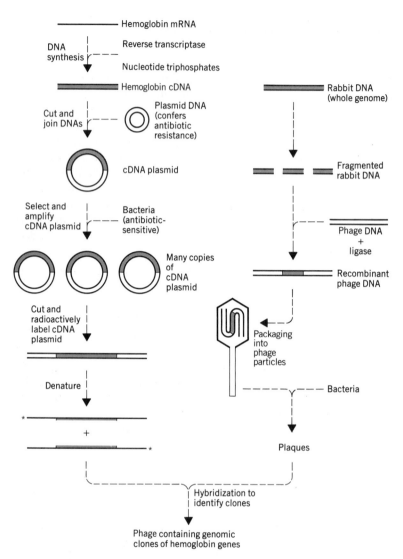

Figure 9-1 Outline of Cloning Procedure for Obtaining Genomic Clones of Hemoglobin DNA. Purified hemoglobin mRNA (*left*) is converted to a DNA form by using the enzyme reverse transcriptase. The DNA copy, called complementary DNA (cDNA), is inserted into a plasmid vector, and bacteria are transformed with recombinant plasmids containing cDNA. Bacteria containing the appropriate recombinants are selected, and large amounts of cDNA plasmid are isolated. This plasmid DNA is cut and radioactively labeled to use as a probe.

purification of rabbit DNA. To make room for the rabbit DNA inside the virus particles, the lambda DNA was shortened (not all the information in lambda DNA is required for the phage to infect bacterial cells). The phage DNA was cut into two pieces by a restriction endonuclease, and fragments lacking information essential for infection were discarded. Rabbit DNA was broken into large pieces and mixed with the phage DNA. Ligase was added to join the DNA fragments. Most of the resulting DNA molecules contained only phage DNA, but often a piece of rabbit DNA had been inserted. The DNA molecules were then coated with phage proteins, neatly packaging the DNA inside phage particles. These phages, which had formed *in vitro* (i.e., in a test tube) were next allowed to infect *E. coli* cells. When the cells were spread on agar, the phages became separated from each other. Phage plaques formed in the lawn of bacteria as each phage reproduced (Figure 6-6). Every plaque arose from a different phage particle, and each plaque contained millions of identical phage particles. Unfortunately, very few contained hemoglobin genes.

At this point the problem was to locate the rare plaques that contained hemoglobin genes. For some genes it is possible to look for specific properties of the plaques arising from recombinant proteins produced by the infected bacterial cells. For example, some proteins catalyze chemical reactions that cause a plaque to become a particular color if the protein is present. Usually this is not the case, however, and nucleic acid probes or specific antibodies must be used to identify plaques that contain the genes being sought. Antibodies, which are useful for identifying the products of specific genes, are described more extensively in a later chapter and are not dealt with further here. In methods using nucleic acid probes, DNA in each plaque is tested for its ability to hybridize with a nucleic acid complementary to the gene or genes being sought, relying on the principle of comple-

---

Rabbit DNA (*right*) is fragmented, and the fragments are joined to bacteriophage DNA. Phage proteins are added to package the recombinant DNA into phage particles. Bacteria are then infected with the recombinant phage, and plaques form. The plaques that contain hemoglobin DNA are identified by hybridization using the radioactive cDNA plasmid as a probe.

mentary base pairing as discussed in Chapter 8 (see Figure 8-1). Generally, nucleic acid probes are radioactively labeled to facilitate detection.

OBTAINING NUCLEIC ACID PROBES

An appropriate nucleic acid probe can be obtained if the protein product of the gene being sought has already been purified. The amino acid sequence for part of the protein is determined, and the genetic code (Figure 2-6) can be used to deduce the sequence of a short DNA sequence likely to be part of the gene. An oligonucleotide having that sequence is then synthesized by an automated procedure in which one nucleotide after another is attached to the end of a growing chain. Inert, particulate resin, already attached to many molecules of the desired first nucleotide (A, T, G, or C), are commercially available. Such a resin is poured into a chromatography column (Figure 3-9), where it can be easily washed by passing solutions through the column. A solution containing many molecules of the second nucleotide in the chain is then added to the column under conditions in which the second nucleotide attaches to the first. Each of the second nucleotide molecules has a group of atoms (a blocking group) attached that prevents the chain from growing longer. The column is then washed free of any second nucleotide that failed to attach to the nucleotide chain. Next the blocking groups are removed from the second nucleotide of each chain so that a third nucleotide can be attached. Then many molecules of the third nucleotide are added to the column, and they attach to the growing chains. As with the second nucleotide, they are blocked on one side, so the chain can grow by only one nucleotide at a time. The process is repeated until the desired length is reached. Commercial laboratories reliably produce oligonucleotides in the length range of 60 to 70 nucleotides. Radioactive phosphorus can be added to an end by incubating the oligonucleotide with an enzyme called a kinase plus radioactive adenosine triphosphate (ATP).

Another source for a radioactive probe is messenger RNA from the gene being sought. In most cases of gene cloning it is very difficult to isolate mRNA representing the information from just a single gene. Generally, cells make many types of messenger at the same time, and the mRNA molecules are chemically and physically too similar to be separated. However, hemoglobin is a special case, for the mRNA in red blood cells is mainly hemoglobin mRNA. Thus in the work we are discussing, rabbit blood was collected, and the red blood cells were concentrated by centrifugation, forming a pellet in the bottom of a test tube. The pellet of cells was resuspended in a small volume of water and salts, and the cells were broken by adding detergents. The resulting lysate was extracted with phenol to help remove proteins, and alcohol was added. Alcohol caused the RNA to form a white precipitate, which was separated from other cell components by centrifugation. The RNA was next dissolved in water and centrifuged in a density gradient. RNA has a unique buoyant density and is readily purified by this method. At this stage the sample still included ribosomal and transfer RNA as well as hemoglobin mRNA. ~~phenol extraction used to remove proteins & ribosomes~~

Hemoglobin mRNA was separated from other types of RNA by taking advantage of a feature unique to many types of mRNA found in higher organisms: these RNAs often have several hundred **adenosines** attached to the 3′ end. A glass column was filled with cellulose to which single-stranded DNA, composed only of **thymidines,** was attached firmly. The RNA mixture was passed through the column. As the RNA percolated through, the stretches of adenosines on the hemoglobin messengers formed complementary base pairs with the long runs of thymidines fixed to the cellulose. The complementary base pairs were strong enough to prevent hemoglobin mRNA from flowing through the column. The other RNA molecules did flow through, and they were discarded. Hemoglobin mRNA was then removed from the column by breaking the base pairs, a step that can be accomplished by warming the column gently.

At this stage the mRNA was almost ready to use to identify a phage plaque containing a hemoglobin gene. But first the information in it, the sequence of nucleotides, had to be converted to a highly radioactive form. This could have been done by mixing RNA, radioactive nucleotides, and a type of DNA polymerase called reverse transcriptase. As pointed out in an earlier chapter, this polymerase is obtained from retrovirus particles, and it makes DNA from free nucleotides, using RNA as a template. The DNA, called complementary DNA (cDNA), can be made highly radioactive if during its synthesis radioactive nucleotides are present in reaction mixtures.

In the initial cloning studies with hemoglobin it was necessary to screen hundreds of thousands of phage plaques, and huge amounts of cDNA were required (as pointed out later, it was not sufficient to clone cDNA because cDNA does not contain the entire nucleotide sequence of the gene). It was decided to first clone hemoglobin cDNA into a plasmid. Then large amounts of the recombinant plasmid could be obtained easily from bacterial cells. This cloned cDNA would be radioactively labeled and used to test the phage plaques for hemoglobin genes. Since the plasmid methods initially used for cloning hemoglobin cDNA are now seldom utilized, a more general strategy is presented in the next two sections to illustrate how biological properties can be used to search large populations of molecules for particular types.

CLONING WITH PLASMIDS

DNA fragments from the organism being examined (in our case cDNA from rabbit) are inserted into plasmid DNA using restriction endonucleases and ligase, the DNA mixture is added to *E. coli* cells, and some of the DNA molecules enter the bacteria. The task is to isolate a pure culture of bacterial cells harboring a plasmid that contains only a particular piece of rabbit cDNA (hemoglobin cDNA). After the bacteria have taken up DNA, they continue to grow and divide, and soon there may be billions of bacteria in the culture. The cultured *E. coli* cells fall into four classes:

1. *E. coli* that failed to take up any plasmid DNA (99.99% of the cells).
2. *E. coli* that took up plasmid DNA without any rabbit cDNA inserted.
3. *E. coli* that took up plasmid DNA with rabbit cDNA inserted but not the particular rabbit gene being sought.
4. *E. coli* that took up plasmid DNA into which the desired rabbit gene was inserted.

The fourth category is the only one of interest, and members of this category are generally very rare.

Two tricks are used to increase the odds for finding plasmids with rabbit genes (classes 3 and 4). First, since the plasmid chosen as a cloning vehicle contains a gene for resistance to **tetracycline,** a drug that prevents bacterial growth, the antibiotic can be added to a flask seeded with the bacteria. This measure, which ensures that only cells protected by the plasmid can grow, eliminates all the cells in category 1. Alternatively, all the cells can be spread onto an agar plate that contains tetracycline. Then only the cells containing a plasmid having a gene for tetracycline resistance will grow and form colonies.

The second trick distinguishes cells containing plasmids joined to rabbit DNA from those that do NOT have a rabbit DNA insert (Figure 9-2). To execute this trick one uses a cloning vehicle, a plasmid, having two genes for antibiotic resistance. Often one gene is for tetracycline resistance (tet^R) and the other for ampicillin (penicillin) resistance (amp^R). Since restriction endonucleases cut in very specific locations, an endonuclease can be found that cuts the plasmid DNA only inside the gene for ampicillin resistance (Figure 9-3). Consequently, whenever rabbit DNA is attached to this plasmid, it will be inserted into the middle of the ampicillin resistance gene, for that is where the ends of the DNA occur. The large stretch of rabbit DNA inserted into this plasmid gene will render the gene inactive. Consequently, cells containing plasmids with rabbit DNA inserted into the ampicillin-resistance gene will be resistant only to tetracycline. On the other hand, cells containing a plasmid having no rabbit DNA will be resistant to both drugs. Thus all the colonies that formed on the tetracycline-containing agar plate (Figure 9-2*d*) can be tested to find ones that *fail* to grow on ampicillin-containing agar.

Large numbers of colonies can be tested by a process called **replica plating.** A piece of sterile velvet is carefully placed on the surface of the tetracycline-containing agar plate so it touches the bacterial colonies. Some of the cells from each colony will stick to the velvet, which is then removed and set onto a clean, ampicillin-containing agar plate. Cells from each colony will drop from the velvet and stick to the agar of the clean plate. There the cells will grow into colonies if they are resistant to ampicillin. The method places cells onto the second plate in a pattern identical to the distribution of colonies on the first plate. Replica plating

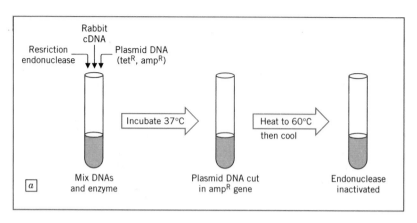

| Resriction endonuclease | Rabbit cDNA | Plasmid DNA (tet^R, amp^R) |

Incubate 37°C

Heat to 60°C then cool

[a] Mix DNAs and enzyme

Plasmid DNA cut in amp^R gene

Endonuclease inactivated

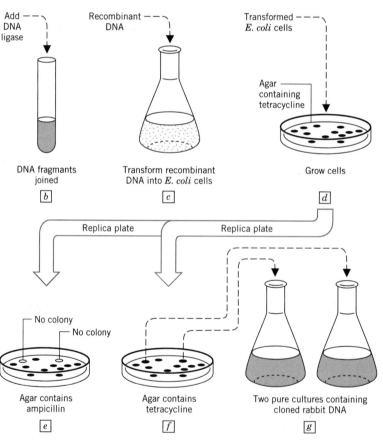

Add DNA ligase

Recombinant DNA

Transformed E. coli cells

Agar containing tetracycline

DNA fragments joined
[b]

Transform recombinant DNA into E. coli cells
[c]

Grow cells
[d]

Replica plate   Replica plate

No colony
No colony

Agar contains ampicillin
[e]

Agar contains tetracycline
[f]

Two pure cultures containing cloned rabbit DNA
[g]

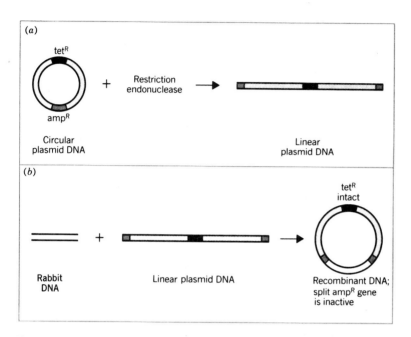

Figure 9-3 Inactivating a Gene. (a) Plasmid DNA containing genes for resistance to ampicillin and to tetracycline is cut with a restriction endonuclease, dividing the *amp*^R gene into two parts. **(b)** When rabbit DNA is inserted into the plasmid, the two parts of the *amp*^R gene remain separated; hence the gene is inactive.

Figure 9-2 Procedure for Obtaining Pure Clones Containing Rabbit DNA. (a) Plasmid DNA conferring resistance to tetracycline and ampicillin is mixed with rabbit cDNA. The DNAs are treated with a restriction endonuclease that cuts the plasmid only once, in the ampicillin resistance gene. The endonuclease is then inactivated by heating the mixture to 60°C. **(b)** DNA ligase is added to join the DNA fragments. The rabbit DNA becomes inserted into the middle of the ampicillin resistance gene, inactivating it (see Figure 9-3). **(c)** *E. coli* cells are transformed with the recombinant DNA. **(d)** Plasmid-containing cells are selected by growth on agar containing tetracycline. Cells with plasmids joined to rabbit cDNA are identified by screening on ampicillin-containing agar **(e)** [these cells grow only on tetracycline **(f)**]. Pure cultures **(g)**, which contain cloned genes, result.

is outlined in Figure 9-2; two colonies are present in Figure 9-2*f* that are absent in Figure 9-2*e*. These colonies contain rabbit DNA inserted into the plasmid.

At this stage one doesn't know which rabbit gene or genes are contained in any particular colony identified as containing rabbit DNA by replica plating. In fact, the odds for obtaining one that has the gene of interest can be very low. Another trick is needed to identify bacterial colonies that have the particular gene being sought. First, a small sample of cells from each colony containing a rabbit DNA fragment (Figure 9-2*f*) is touched with a sterile toothpick and transferred to a new agar plate, where a gridlike pattern of colonies will be set up (Figure 9-4). Sometimes a cloned gene will give its bacterial host a selective growth advantage; in such cases the components of the agar are adjusted to permit growth only of colonies that contain the gene being sought. Genes from complex organisms, such as rabbits or humans, rarely offer such a shortcut, however. Usually,

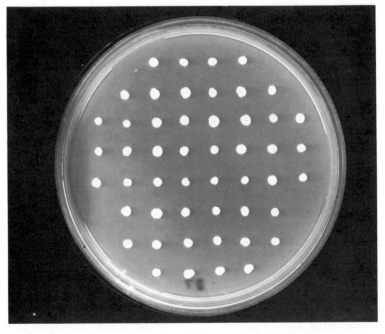

Figure 9-4 Agar Plate Bacterial Colonies Arranged in a Grid Pattern

the appropriate colony is located by nucleic acid hybridization or by reaction of the colony with antibodies specific to the protein product of the gene being sought.

SCREENING BY NUCLEIC ACID HYBRIDIZATION

Screening bacterial colonies for a particular gene is carried out as described in Chapter 8. Briefly, colonies are grown on an agar plate and transferred to a piece of filter paper in the same grid pattern they had held on the agar (sometimes the cells are grown directly on filter paper placed on the agar surface); then the paper is transferred to a dilute solution of **sodium hydroxide** (lye or caustic soda). The sodium hydroxide breaks the cells, and some of the cellular debris plus the cellular DNA stick tightly to the paper. The sodium hydroxide also causes the DNA to become single-stranded. Since the bacterial colonies had been arranged in a grid on the paper, DNA released from the cells forms the same pattern. Next, the sodium hydroxide is neutralized with acid, and the paper is slipped into a dish containing a radioactive probe. This probe is often made synthetically from information derived from the amino acid sequence of the protein, as described earlier in this chapter.

In the case of hemoglobin gene cloning, a small amount of hemoglobin cDNA was available which was cloned into a plasmid to prepare many copies. The cDNA also served as a template for preparing a radioactive RNA probe by incubation with RNA polymerase.

As pointed out in Chapter 8, both the radioactive probe and the cellular DNA attached to the paper are single-stranded, ready to form base pairs. The probe will form base pairs with paper-bound DNA only if the paper-bound DNA contains the gene being sought (see Figure 8-1). When the target gene is present on the filter, the radioactive DNA will be indirectly bound to the paper, and the location of the radioactivity will identify the bacterial colony that contains the rabbit gene being sought.

To determine where the radioactivity is located, the paper is removed from the dish and washed thoroughly to remove any radioactive probe that is not base-paired with paper-bound DNA. X-ray film is then placed next to the paper (Figure 9-5*a*). Wherever radioactive probe is base-paired to paper-bound DNA, it exposes the film, producing a dark spot (Figure 9-5*b*). These dark spots correspond to bacterial colonies containing the gene of interest. This searching procedure makes it possible to obtain a pure culture of *E. coli* in which each cell contains a plasmid into which the gene of interest has been inserted. Those cells can then be grown to huge numbers, from which many recombinant plasmid DNA molecules can be isolated. Large amounts of hemoglobin cDNA were obtained by purification from such bacterial cells.

The hemoglobin cDNA was then used to screen the lambda phage plaques for those containing rabbit hemoglobin genes. The cDNA plasmids were linearized by cutting with a restriction endonuclease and radioactively labeled using an enzyme that could add a radioactive phosphorus atom to the end of DNA. This generated large amounts of a radioactive hemoglobin probe. Then more than 750,000 phage plaques were examined, leading to the discovery of four that contained hemoglobin genes. These clones, obtained directly from rabbit DNA, are called **genomic clones** to distinguish them from the cDNA clones derived from mRNA. This distinction is important, for as we shall see in Chapter 11, mRNA from higher organisms often contains only parts of the nucleotide sequence of the gene. Thus cDNA clones would also contain only parts of the gene.

PERSPECTIVE

Often existing technology dictates the direction taken by research. For example, in the mid-1970s methods were available for isolating hemoglobin mRNA from red blood cells, the cells specialized to make hemoglobin. These mRNA molecules could be used as templates for the synthesis of radioactive probes to locate cloned hemoglobin genes. This is a rare situation, since most genes are transcribed into RNA in cells that make mRNA of so many different types that it is prohibitively difficult to separate them from each other to make hybridization probes for particular genes. Thus although other genes might have been just as interesting, hemoglobin genes were an obvious choice for these early cloning studies.

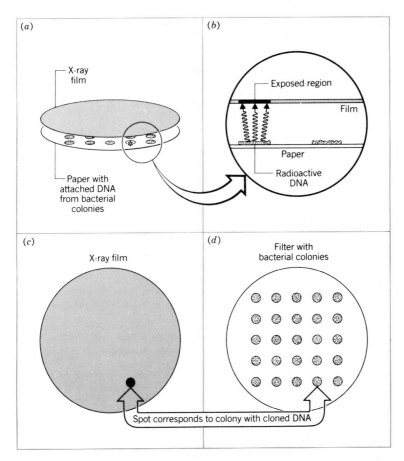

Figure 9-5 Detection of Bacterial Colonies Containing a Particular Rabbit Gene. Bacterial cultures are grown on paper placed on the surface of an agar plate (nutrients seep through the paper). The paper is removed and is dipped in sodium hydroxide to break open the cells in the colonies and denature the DNA. The denatured (single-stranded) DNA becomes attached to the paper. The sodium hydroxide is neutralized, and radioactive probe DNA is added. Complementary base pairs form between the probe DNA and paper-attached DNA if the gene being sought is present in the bacterial colony . The paper is washed and placed next to X-ray film **(a).** If a particular bacterial colony (*) contain the gene of interest, its DNA will have hybridized to the radioactive probe, and the X-ray film will be exposed above the colony **(b).** The pattern of exposed spots on the film **(c)** is used to identify the cultures containing cloned genes by comparison with the original distribution of colonies on filter paper **(d).**

The cloning strategies themselves also illustrate how biological research builds on previous developments. Messenger RNA isolation, phage and plasmid manipulation, restriction endonuclease cutting of DNA, enzymatic synthesis of DNA, and nucleic acid hybridization were all highly refined technologies being used for other studies before they were applied to gene cloning. Gene cloners combined these methodologies in a new way to serve a new purpose. Indeed, most progress in science comes from combining selected aspects of existing knowledge.

Questions for Discussion

1. What information do you need to have about a gene to clone it?
2. Suppose you wish to clone the gene that codes for tetracycline resistance. Devise a cloning procedure that would require no radioactive probes.
3. If the amino acid sequence of a protein is known, it is sometimes possible to use the genetic code (Figure 2-6) to help clone the gene. This is done by deducing the nucleotide sequence for a short region of the gene from the amino acid sequence of the protein. Next an oligonucleotide (usually about 15 nucleotides long) is synthesized that has the nucleotide sequence predicted from the amino acid sequence of the protein. This oligonucleotide is then radioactively labeled and used as a probe for detecting bacterial colonies or phage plaques that contain the gene. The most suitable oligonucleotide would be constructed from a region of the protein that would be rich in which amino acids? Regions rich in which amino acids would be least suitable?
4. Suppose you wish to clone a gene encoding a protein whose amino acid sequence is not known, but you do have antibodies that will bind specifically to the protein and cause it to precipitate. (If you are unfamiliar with antibodies, see Chapter 11.) How could the antibodies be used to obtain the messenger RNA

from the gene encoding the protein? (Hint: See frontispiece of Chapter 4.)

5. Oligonucleotides of specified sequence can be obtained commercially. If the technology limits the length to 60 nucleotides, how would you construct a piece of DNA having a specified sequence 300 nucleotides long?

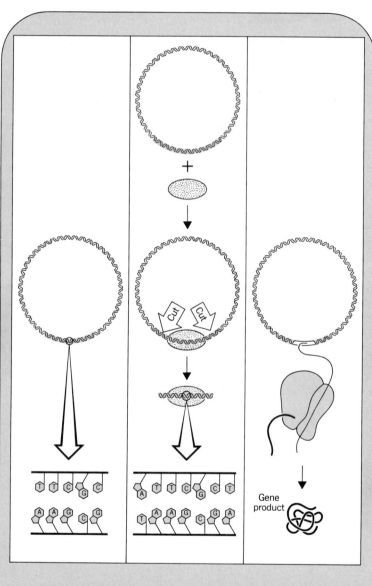

Cloned genes are used to determine nucleotide sequences,
to identify sites where proteins bind to DNA, and to produce
large amounts of gene product.

C H A P T E R   T E N

CLONED GENES AS RESEARCH TOOLS
Gene Structure, Expression Vectors, and
Gene Function

Overview

Gene cloning technologies make it possible to obtain the large amounts of specific regions of DNA necessary to study gene structure and function. Gene structure is analyzed by determining the nucleotide sequence of a section of DNA. To do this, DNA polymerase is used to produce a series of DNA fragments that extend from a specific spot in DNA to particular nucleotides in the sequence being analyzed. By measuring the lengths of these fragments and by knowing the identity of the last nucleotide added by DNA polymerase in each fragment, it is possible to determine the position of each nucleotide relative to the fixed spot. The nucleotide sequence can then be deduced. The exact position of a gene can be located by finding a nucleotide sequence identical to that predicted from the amino acid sequence of the protein product of the gene.

Gene function is studied by examining (1) how alterations (mutations) in the gene affect the life of the organism under study and (2) how the purified protein product of a gene interacts with other proteins, with DNA,

179

and with small molecules. Gene cloning makes it possible to generate specific mutations and to produce the large amounts of particular proteins needed for this type of study.

INTRODUCTION

We have now introduced the basic principles of molecular genetics, described the tools used to clone genes, and sketched a complex example of gene cloning. It may have appeared that cloning a particular gene is an end unto itself. It is not. Gene cloning is but a tool, albeit a very powerful tool, that biologists use to explore the molecular properties of life. Through gene cloning we can obtain specific DNA fragments and particular protein molecules in large enough quantities to study gene structure, function, and regulation.

The easiest information to gain from gene cloning is the nucleotide sequence of the DNA. Indeed, there are now commercial laboratories that specialize in sequence analyses. Since the genetic code is the same for all life-forms on Earth, the nucleotide sequence is easily converted into amino acid sequences. At that point many options appear. The cloned genes can be used to clone related genes or to direct the synthesis of large amounts of their protein products. Clones also can be altered and placed in organisms to study how defective forms alter the lives of their hosts. The aim of this chapter is to provide a general sense of how cloning is contributing to our understanding of life. Later chapters will elaborate on the contributions.

ANALYSIS OF GENE STRUCTURE

The initial step in analyzing gene structure is to determine the nucleotide sequence of the gene. DNA molecules are incredibly long and chemically monotonous; deciphering the exact order of thousands of As, Ts, Gs, and Cs requires an ingenious combination of many of the molecular tools discussed in earlier chapters.

The recipe outlined in Chapter 9 makes it possible to insert a particular DNA fragment into either plasmid DNA or phage DNA. The resulting recombinant DNA is copied trillions of times by bacteria, and these molecules are purified. Further analysis sometimes requires that the cloned DNA, the insert, be separated from the cloning vehicle (the plasmid or phage DNA). The first step is to free the insert. This is usually accomplished by cutting the recombinant DNA with restriction endonucleases. If the cloned DNA fragment had been inserted into the cloning vehicle at a site where a specific restriction endonuclease cuts and if the joining event regenerated the recognition site for the enzyme (Figure 10-1), the recombinant DNA need only be treated with the same restriction endonuclease

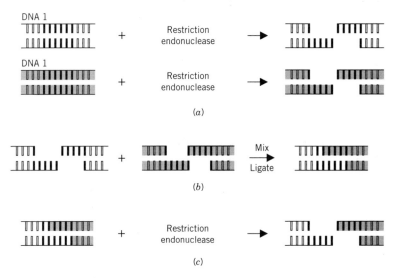

Figure 10-1 Regeneration of a Restriction Site (a) Two DNA molecules containing a recognition site for the same restriction endonuclease (solid regions) are cut with that enzyme. When the DNA molecules are ligated together **(b)**, the restriction site is regenerated. The ligated DNA can in turn be recut into fragments **(c)** by subsequent treatment with the restriction endonuclease. This feature makes it possible to insert a DNA fragment into a plasmid and later excise the fragment by cutting with the same restriction endonuclease.

to liberate the insert from the cloning vehicle (see Figure 7-6). If this is not the case, restriction endonucleases must be found that cut outside the insert. The second step is to physically separate the insert DNA from the cloning vehicle. Since the two usually have different lengths, they can be separated into discrete bands by gel electrophoresis (see Figures 7-4 and 10-2). DNA from a particular band can be recovered by slicing the gel where the band is located, transferring the slice to a second electrophoresis apparatus, and then applying

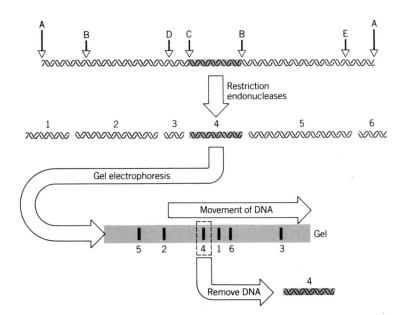

Figure 10-2 Cutting DNA into Pieces of Manageable Size. A cloned human DNA fragment is often much longer than the region being sought (shaded region between **C** and **B**); thus, it must be cut into smaller pieces. Arrows **A** represent the ends of the human DNA fragment, and arrows **B, C, D,** and **E** indicate cleavage sites for four different restriction endonucleases. Treatment of the human DNA fragment with these four nucleases produces fragments **1** through **6**; these fragments are physically separated by gel electrophoresis. Each band in the gel (Figure 7-4) results from billions of identical DNA molecules. Portions of the gel can be excised, and the DNA can be removed from the gel. In this example, band **4** would be saved for further study.

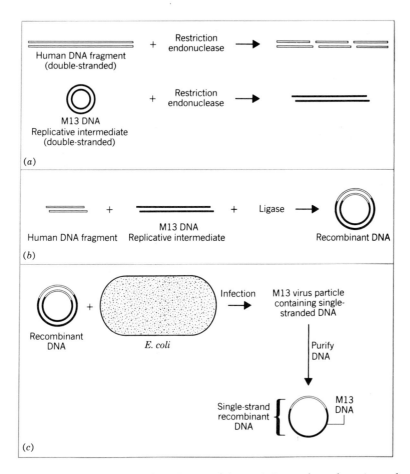

Figure 10-3 Cloning into Phage M13. (a) A restriction endonuclease is used to cut a human DNA fragment and a purified double-stranded replicative intermediate of phage M13 (single-stranded DNA viruses form double-stranded DNA molecules as a part of their life cycle). (b) The DNAs are mixed and ligated to form a recombinant DNA. (c) The recombinant DNA is used to infect *E. coli* cells, producing M13 virus. The virus particles contain only one of the two DNA strands. This DNA strand, which is the same for all the virus particles, can then be purified in large quantities.

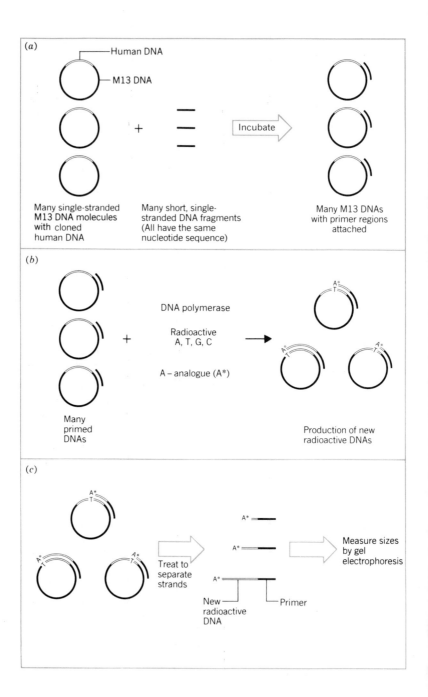

an electric current to drive the DNA out of the gel slice. If necessary, the DNA fragments can be cut into still smaller fragments with other restriction endonucleases, and these new fragments can be further separated by gel electrophoresis or recloned by insertion into a plasmid or phage DNA. Thus the DNA fragments can be gradually pared down to obtain pieces small enough for nucleotide sequence analysis.

Although several methods have been developed to determine nucleotide sequences, all use the same general strategy: a reference point is established, and the distance is measured from the reference point to each of the nucleotides of a given type—for example, to each A, to each G, to each C, or to each T. In one method, the DNA fragment under study is first cloned into a specific site in the DNA of a bacteriophage called M13. During infection of bacteria, this particular phage packages only one of its DNA strands; thus, recombinant DNA isolated from these virus particles contains only one of the two DNA strands (see Figure 10-3).

To establish a reference point, the DNA is mixed with a short, single-stranded piece of DNA that is complementary to a region of M13 DNA near the position where the gene to be sequenced has been inserted. The short piece forms base pairs with the M13 DNA, creating a short, double-stranded region of DNA that can serve as a primer for DNA synthesis (Figure 10-4*a*). A large number of these partially double-stranded DNA molecules are mixed with DNA polymerase and recombinant M13 template. DNA polymerase syn-

---

Figure 10-4 Creating a Collection of Variable Length Copies of Cloned DNA. (a) Many single-stranded M13 recombinant DNA molecules from Figure 10-3 are mixed with short single-stranded DNA fragments complementary to a region of M13 DNA near the junction between M13 DNA and the cloned DNA. The fragment forms a double-stranded region with M13 DNA. (b) The short double-stranded region acts as a primer for DNA polymerase. New DNA is synthesized by adding radioactive As, Ts, Gs, and Cs onto one end of the primer. An analogue of A (A*) is added to halt synthesis at the various positions where its complement, T, occurs in the nucleotide sequence of the cloned DNA. (c) The DNAs are treated to separate the strands, and the lengths of the single-stranded molecules are measured by gel electrophoresis.

thesizes new DNA from the template, beginning at one end of the short, double-stranded region (DNA polymerase requires a primer to start synthesis of DNA; thus, its synthesis begins *only* where the primer is located). DNA polymerase soon crosses into the cloned DNA region and uses it as a template to make new DNA. The polymerase can be stopped by including in the reaction a nucleotide analogue (a fake nucleotide) that lacks a 3′ oxygen (see Figure 2-2 for nucleotide structure). Such a nucleotide cannot be part of a DNA backbone and will not allow the new DNA chain to grow beyond where the analogue is inserted. If the analogue behaves like an A, the new DNA chain will stop after it would normally have added an A (Figure 10-4*b*). Normal As are included in the reaction mixture to allow some DNA synthesis before an A analogue stops the reaction. Since there are many As in a nucleotide sequence, the stops will not always be in the same place. Thus, by carefully adjusting the amount of analogue in the reaction mixture, it is possible to create a collection of DNA molecules that begin at a distinct spot (the end of the primer) and stop at the various positions where As occur.

The procedure just described is repeated in three other test tubes using the respective analogues for G, C, and T. Thus four separate collections of molecules are made. Every molecule starts at the same place, and the various types end at different distances from the starting point. The fragments are then radioactively labeled at an end so that they can be detected. Figure 10-5 illustrates such a collection in which the reaction was stopped by an analogue of A.

The lengths of the newly synthesized molecules are measured by gel electrophoresis under conditions in which the DNA molecules are in a single-stranded configuration (Figure 10-5). The four collections of molecules (terminated at As, Gs, Cs, or Ts) are subjected to electrophoresis in lanes next to each other, and after electrophoresis a piece of film is exposed by the radioactivity in the DNA. By examining only radioactive molecules, it is not necessary to consider the many nonradioactive DNA fragments that might otherwise complicate the experiments. A series of bands is seen (Figure 10-6; the electrophoresis methods are precise enough to permit the separation of DNA molecules differing by only one nucleotide). The lowest band in Figure 10-6 represents molecules that extend to an A, for that band appears in the sample containing the A analogue (A*).

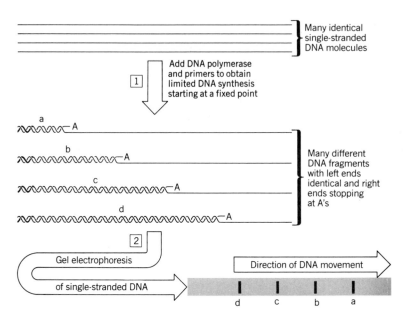

Figure 10-5 Creating and Analyzing a Collection of DNA Fragments Having Different Lengths. (1) Many identical single-stranded DNA molecules are partially replicated as in Figure 10-4; a collection of shorter molecules (a–d) is formed. All the new DNAs have the same left end but different right ends, which in this example always stop at an **A.** (2) The collection of fragments is denatured (converted into single strands) and analyzed by gel electrophoresis to determine the length from the left end to the position of each **A** (the distance a DNA molecule moves in the gel during electrophoresis is related to its length).

The next higher band is in the G-analogue lane, so the next nucleotide is a G. Thus nucleotide sequences can be read directly from pictures such as that shown in Figure 10-6.

The nucleotide sequences from a number of adjacent restriction fragments can be fit together to produce very long sequences. Particular genes are located in the sequence by comparing the nucleotide sequence to that expected from the amino acid sequence of the protein made from the gene. For example, if the left-most amino acid in the protein is methionine, the second one from the left is tryptophan, and the third one is phenylalanine, we expect the

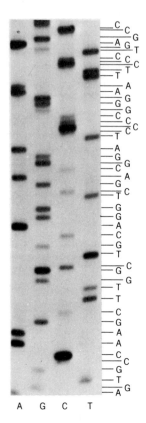

Figure 10-6 DNA Sequencing Gel. Four radioactive DNA samples were subjected to electrophoresis in adjacent lanes of a gel. The four samples, labeled A, T, C, and G across the bottom of the figure, were synthesized in the presence of analogues to A, T, C, and G, respectively, as described in Figure 10-4. The nucleotide sequence of the DNA is indicated on the right. (Photograph courtesy of Richard Archer, Janice Zengel, and Lasse Lindahl, University of Rochester.)

nucleotide sequence of that portion of the gene to be ATGTGGTT (T or C) because we know the triplet encoding each amino acid (see Figure 2-6). The ninth nucleotide could be either T or C because phenylalanine is encoded by two triplets, TTT and TTC. This type of comparison allows us to determine the exact position of a gene. Nucleotide sequences can also be scanned for regions capable of encoding a protein. These regions, called **open reading frames** (ORFs), contain long stretches of nucleotides between start and stop codons, and they frequently signify the location of genes thus far unidentified. Control regions can also be spotted when their sequences are similar to those known to control the expression of other genes. Thus nucleotide sequence analysis is a major tool in studying DNA anatomy.

CLONING MORE GENES

As pointed out earlier, the nucleotide sequence of the DNA of any organism reflects the evolutionary history of the organism. As changes accumulate in DNA that allow organisms to adapt to particular ecological niches, genes responsible for the same process will tend to diverge, to have different sequences. This creates families of organisms. Within an organism, duplications of short regions occasionally occur in DNA, and over generations the two copies accumulate sequence changes that eventually allow their protein products to have different, but sometimes related functions. This creates families of genes. Within families of organisms and families of genes there sometimes remain regions of the genes that have changed little. Consequently, a gene cloned from one species often can be used as a probe to clone the homologous gene from another organism. Likewise, if the cloned gene is a member of a family of genes, it can be used as a probe to pull out the other members of the family.

The ability to cross species lines allows biologists to study fundamental processes in microorganisms and then apply that work directly to humans. Repair of errors in DNA provides an example. All organisms contain genes whose protein products are responsible for correcting the nucleotide sequence of DNA when the wrong nucleotide has been inserted by DNA synthesis. Extensive study of these genes in *E. coli* and yeast led to the observation that they are members of a family of genes. It seemed reasonable that human DNA would also contain members of the family. An examination of DNA damage in mutant yeast cells that were defective in ability to repair DNA showed a remarkable likeness to the damage seen in certain human cancers. It turned out that a human repair gene, cloned using a yeast nucleotide sequence as a probe, was a gene involved in a type of colon cancer. This observation contributed to the general concept that certain forms of cancer correlate with the accumulation of DNA damage.

EXPRESSION VECTORS

Since proteins, not DNA, do the work in cells, biologists often use the cloned gene to obtain large amounts of its protein product. One way to accomplish this is to move the cloned gene into a specially constructed plasmid called an **expression vector** (see Figure 10-7). The expression vector contains an easily controlled promoter, such as that from the *lac* operon (see Chapter 4), situated upstream from a series of closely spaced sites for restriction endonucleases. Thus the expression vector plasmid can be easily cut by a variety of nucleases immediately downstream from the promoter. Once the general features of the cloned gene (i.e., the location of the beginning and end of the gene) have been determined, the cloned gene can be trimmed and tailored to permit insertion into the expression vector next to the promoter. Bacterial cells are then transformed with the expression vector, and when the inducer of the promoter is added to the cell culture, the bacterial cells will begin to produce protein from the cloned gene. Systems have been developed in which 40% of the cellular protein is made from the cloned gene.

Bacterial systems are not always suitable for high level production of eukaryotic proteins. One problem arises from glycosylation, the process in which sugars are added to newly synthesized proteins. In higher organisms glycosylation can be quite important for protein activity, for directing the protein to the proper location in the body, and for protecting it along the way. Bacteria do not glycosylate proteins. Another problem occurs when high level expression leads to formation of large, intracellular protein aggregates, since it is sometimes difficult to obtain active protein from the aggregates. These and other problems can be bypassed with expression systems that use yeast or mouse cells.

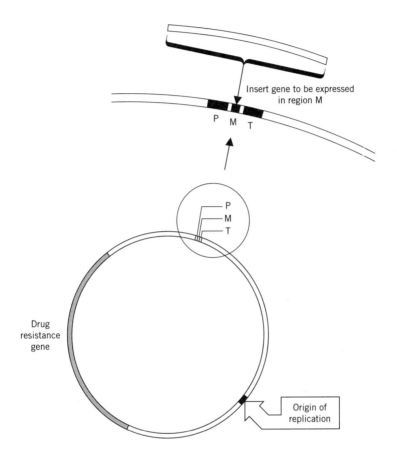

Figure 10-7 A Plasmid Expression Vector. In this circular map of a plasmid, an inducible promoter, labeled **P,** is followed closely by a multiple cloning site **M** (a short region of about 40 base pairs constructed to allow many different restriction endonucleases to cleave the DNA there). Farther downstream is a DNA sequence that serves as a terminator **(T)** to stop transcription. A gene of choice is placed in the plasmid at the multicloning site. The plasmid also contains an origin of replication (to allow it to multiply inside bacteria) and a drug resistance marker (to enable the biologist to select for bacteria that harbor the plasmid). Promoters are used which are controlled by a repressor that can be easily regulated by additions to the growth media of the cells; thus production of the protein of choice can be induced whenever desired.

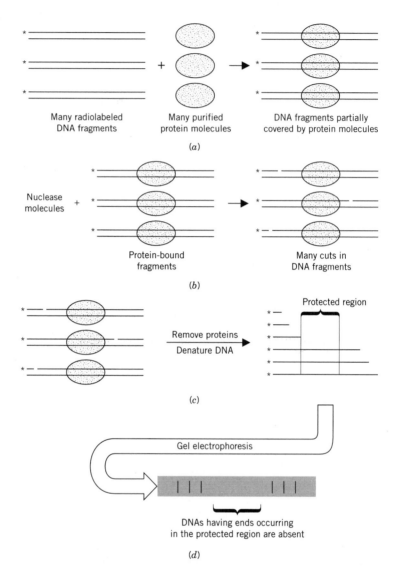

Many radiolabeled
DNA fragments

Many purified
protein molecules

DNA fragments partially
covered by protein molecules

(a)

Nuclease
molecules

Protein-bound
fragments

Many cuts in
DNA fragments

(b)

Remove proteins
Denature DNA

Protected region

(c)

Gel electrophoresis

DNAs having ends occurring
in the protected region are absent

(d)

Figure 10-8 Protection of DNA by a Protein. (a) DNA fragments that have a radioactive label (asterisk) on one end of one strand are mixed with purified protein molecules. The proteins bind to the DNA fragments, covering a short region of DNA. **(b)** Nuclease molecules are mixed with the protein–DNA complexes, and the DNA strands are cut. Cutting can occur anywhere except where the protein is bound to the DNA. **(c)** The proteins are removed by detergent and protease treatments, and the double-stranded molecules are converted by mild heat or alkali treatment into single-stranded ones.

ANALYSIS OF GENE FUNCTION

In many studies the central questions revolve around what particular genes do and how they do it. One of the general approaches for learning about gene function is to perturb the gene inside cells and then see what happens. This is often done by creating mutant genes, genes whose nucleotide sequence has been altered. For example, DNA molecules can be delivered to mouse embryo cells that develop into mice, yielding fat mice, skinny mice, giant mice, and cancerous mice for study purposes. Conclusions are then drawn about the behavior of comparable genes in humans.

In another general approach, an expression vector is used to produce the protein product of a gene in large amounts. Then the protein can be purified and its interactions with other pure proteins, DNA, or small molecules can be studied. For example, if you think that two proteins might act by binding tightly together, you can determine whether the two purified proteins do this. Gene cloning comes into play because many of the important regulatory proteins are present in living cells in very small amounts, making it difficult to obtain enough of a particular protein to study its properties and interactions with other molecules. In a number of cases gene cloning and the use of expression vectors has erased the problem.

Some of these protein products are particularly interesting because they act on DNA. Repressors are an example. Chapter 4 pointed out that repressors prevent RNA polymerase from transcribing RNA from certain genes, and thus repressors regulate gene expression. Understanding this aspect of gene control involves knowing exactly where the repressor binds to the DNA. Gene cloning technologies

---

The nuclease treatment in (b) causes the DNAs to have many different lengths; only radioactive DNAs are considered because only radioactive ones are measured below in (d). Note that not all DNA sizes are present because the protein blocks the cutting in certain regions during the nuclease treatment in (b); the protein produces a protected region of DNA. **(d)** The lengths of the radioactive DNAs are analyzed by gel electrophoresis; no bands are observed that have ends occurring in the protected region. In a sense, the protein has left its footprint on the DNA, and this type of analysis is often called **DNA footprinting.**

have made it possible to obtain large amounts of both repressor protein and the potential DNA binding sites. When the two are mixed, complexes form between the protein and the DNA. DNA in these complexes is protected from cleavage otherwise produced when nucleases are added to the preparation (Figure 10-8). Consequently, analysis of nucleotide sequences that survive nuclease treatment provides insight into repressor binding sites, and this information adds to our general understanding of gene organization and regulation (see Figure 4-7).

Protein engineering can also be carried out by inserting specific changes into a gene and then expressing large amounts of the mutant protein in bacteria. After the protein has been purified, its structure can be studied to see how changing a particular amino acid alters the structure and chemistry of the protein.

PERSPECTIVE

We know in general terms how DNA acts as a repository for information. By determining the nucleotide sequences of DNA molecules, we are rapidly learning exactly what that information is. But this new knowledge will not suddenly tell us how life works. Fundamental problems remain to be solved even with organisms as simple as bacteria. For example, knowing the exact order of all 4,400,000 base pairs in *E. coli* DNA will not tell us how the timing of cell division is controlled. To understand cell division, and most other cellular processes, we must know how the products of the relevant genes work and interact. Gene cloning makes it much easier to apply both biochemical and genetic techniques to a single problem. Genetics utilizes mutations to perturb the normal function of a gene in a living cell. As the effects of the mutations are examined, ideas develop about how the genes and their products interact. For example, when a mutation in a particular gene causes DNA synthesis to stop, we know that the product of that gene is somehow involved in DNA synthesis. Gene cloning allows us to purify large amounts of the gene product so that its interactions with other molecules can be studied biochemically. The next two chapters discuss results derived from the combination of gene cloning, genetics, and biochemistry.

Questions for Discussion

1. DNA fragments from all parts of the *E. coli* chromosome have been cloned into bacteriophage lambda, and a large collection of these recombinant phages is available. For each cloned fragment the position on the chromosome is known. How could you use this collection of phages to set up a system to determine the map location of any *E. coli* DNA fragment that you happened to clone?

2. RNA molecules can be synthesized that have a nucleotide sequence complementary to a given messenger RNA. Such a complementary RNA molecule is called **antisense RNA**. How might antisense RNA be useful in fighting virus infections? What information would you need to be able to design an antisense RNA?

3. Suppose you were studying antibiotic resistance in the bacterium that causes tuberculosis. The organism is difficult and dangerous to grow, but DNA can be obtained from drug-resistant strains. If you know the nucleotide sequence of a gene in which mutations are often associated with a particular type of antibiotic resistance, how could you use the polymerase chain reaction (PCR) and nucleotide sequencing methods to determine the exact nucleotides altered in particular strains you isolated?

4. When a biologist includes a newly determined nucleotide sequence in an article submitted for publication, most journal editors insist that sequence data for both DNA strands be determined. Why?

5. Gel electrophoresis is used to determine DNA sequences by separating DNA fragments based on length. In general, the sequence of only 200 to 300 nucleotides can be accurately determined. Why is there a size limit? (Hint: Examine the spacing between bands in Figure 10-6.)

6. Commercial laboratories routinely determine DNA sequences from DNA fragments amplified by PCR and supplied by the customer. The customer must also supply short oligonucleotides as primers. Why are the primers needed?

THE NEW
MOLECULAR GENETICS

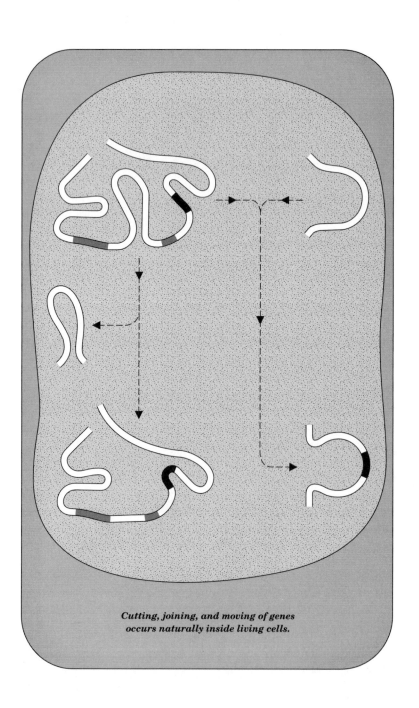

Cutting, joining, and moving of genes occurs naturally inside living cells.

C H A P T E R   E L E V E N

BEYOND THE
CENTRAL DOGMA

A Sampling of Insights Derived from
Gene Cloning

Overview

Gene cloning technologies have allowed us to establish the precise locations of many genes by determining the nucleotide sequences of regions containing the genes. These studies have confirmed that the general biochemical principles of life are the same in humans and in bacteria. However, some remarkable differences exist. Unlike bacterial DNA, our DNA contains nonfunctional copies of genes. Moreover, our functional genes contain long stretches of nucleotides that code for nothing. They are excised from transcripts before they are translated into protein. The sequencing studies have also demonstrated that cutting, moving, and joining specific sections of DNA are normal processes that occur inside living cells.

INTRODUCTION

The central dogma of molecular biology states that genetic information is stored in DNA and flows through RNA to protein. The first four chapters described elements of the central dogma and the next six outlined their application to the process of gene cloning. Some of the fruits of gene cloning can now be described. Six examples have been chosen for this chapter to illustrate how dynamic nucleic acids are. The first involves the globin gene family and serves to introduce the concept of **pseudogenes.** Pseudogenes appear to be relics of once-active genes, and their existence raises questions about gene duplication and evolution. The second and third cases concern RNA splicing. It is now clear that RNA can cut and splice itself; therefore enzymes may not always be protein molecules. The fourth case, antibody formation, illustrates how gene rearrangements play a key role in the development of immunity. The fifth case focuses on transposition, the movement of small parasitic stretches of DNA from one region of DNA to another. The sixth describes an intimate relationship between a bacterium and a plant cell in which a portion of a bacterial plasmid is transferred to the plant genome, producing a tumor and also providing a way to engineer plants.

HEMOGLOBIN GENES AND PSEUDOGENES

The blood protein **hemoglobin** has been extensively studied for many years, and a number of statements can be made about it. First, hemoglobin is composed of four subunits, four separate protein chains called **globins,** that spontaneously associate to form the active protein. The four separate protein chains are of two types, alpha (α) and beta (β). The two types of protein, which differ slightly in length and in amino acid sequence, are paired in the hemoglobin molecule. Thus the predominant adult hemoglobin is generally called $\alpha_2\beta_2$ (Figure 11-1). Second, several types of hemoglobin exist, and at

Figure 11-1 Structure of Hemoglobin. A molecule of hemoglobin is composed of two each of two types of protein subunit. In the predominant form of adult hemoglobin, these subunits are called alpha (α) and beta (β).

different stages of life our genes instruct our blood cells to produce different globin proteins (see Figure 11-2). Each kind of hemoglobin is distinguished by having subunit chains of different types. For example, the blood of young human **embryos** contains two types of embryonic hemoglobin, $\alpha_2\varepsilon_2$ and $\zeta_2\varepsilon_2$. After 8 weeks of gestation, the embryonic forms are gradually replaced by the **fetal** form of hemoglobin, $\alpha_2\,\gamma_2$. Fetal hemoglobin, which predominates until about 6 months after birth, is replaced by the adult forms, $\alpha_2\delta_2$ and $\alpha_2\beta_2$. A third statement is that separate genes code for each of the hemoglobin subunits, ζ, α, ε, γ, δ, and β. Thus we are faced with the question of how the various genes are switched on and off during our development to produce the correct types of hemoglobin for each stage.

Recombinant DNA technologies and DNA sequencing studies have not yet disclosed how gene switching occurs, but they do allow us to make four more statements about how the genes are organized (see Figure 11-3 for schematic drawing of globin gene arrangement). First, globin genes fall into two classes. The α class includes genes α and ζ, while the β class includes genes β, γ (subtypes G and A), ε, and δ. Second, the genes in one class are located in the same region of DNA; but members of the other class are located far away on another chromosome. Third, the proteins in one class, and thus the genes that code for them, have similar structures. For example,

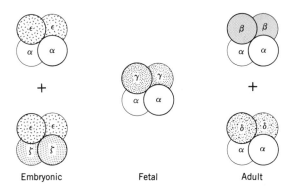

Embryonic   Fetal   Adult

Figure 11-2 Hemoglobin Changes During Development. As humans develop from embryo to adult, changes occur in the protein content (indicated by Greek letters) of their hemoglobin.

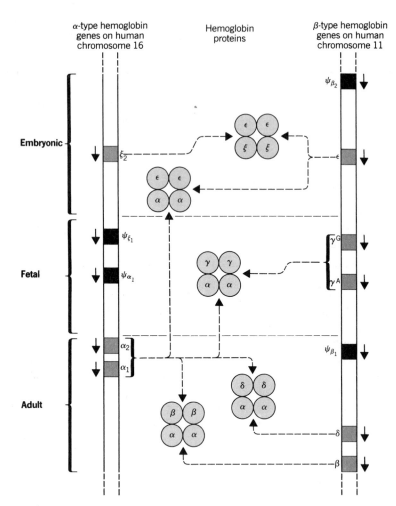

Figure 11-3 Human Globin Genes and Their Protein Products.
Hemoglobin is composed of two types of protein, and the composition
varies with the stage of development through differential activation of globin
genes. The embryonic forms predominate until 8 weeks of gestation, the
fetal form to 6 months after birth, and the adult forms from 6 months on.
Each set of four circles represents a hemoglobin molecule; solid arrows
indicate the direction of transcription, and solid areas represent pseudogenes
(nonfunctional genes; see text). The β-type genes are scattered over a stretch
of 52,000 nucleotide pairs; the α-type genes fall within a region containing
36,000 nucleotide pairs.

all the protein products from the embryonic and adult α-class genes are 141 amino acids long and vary only slightly in amino acid sequence. Fourth, the human genes in a class map in the same order in which they are expressed during development. For example, the β-class genes map in order ε (embryonic), γ^G (fetal), γ^A (fetal), δ (adult), and β (adult) (Figure 11-3). Order is also preserved in the direction of synthesis of messenger RNA from the genes. For each gene, RNA synthesis starts at the end closest to the embryonic gene (arrows, Figure 11-3), and all the messenger RNAs are made from the same strand of DNA (the two DNA strands are complementary, not identical, and they would not code for the same proteins). Why this ordering exists may become clearer when we understand how the developmental switches in globin gene expression occur.

Sequencing studies of the globin gene regions also uncovered several other short stretches of DNA that look remarkably like bona fide globin genes. Careful analysis of the nucleotide arrangements showed that these "genes" (solid regions, ψ in Figure 11-3) are incapable of producing functional globin proteins—they are full of mutations and abnormalities. Premature stop codons, frameshift mutations, abnormal RNA polymerase binding sites (promoters), faulty initiation codons, and large internal deletions ensure that no functional globin proteins can come from these genelike regions called pseudogenes. The study of the hemoglobin genes and the surprising discovery of pseudogenes confronts us with a new set of questions. Did pseudogenes arise from the duplication of a preexisting gene? Did all the modern globin genes arise by duplication of a primitive ancestor? How does gene duplication occur?

The hemoglobin system is complicated, and, as in all complicated biological systems, there are many steps at which things can go wrong and produce serious disease. One class of disease arises from nucleotide substitutions in the genes, causing amino acid changes in the globin proteins. Of the 300 or so globin mutations that have been identified, the most widely known is the mutation in the β-globin gene that causes sickle-cell disease (see Figure 3-5). Another group of hemoglobin disorders, called the thalassemias, comes from a deficiency of one specific type of globin protein. Still other problems arise when the normal switch from one form of hemoglobin to another fails to occur.

EXONS, INTRONS, AND RNA SPLICING

Once genes had been purified, it was a straightforward process to determine the nucleotide sequences of the genes and of the DNA surrounding them. This led to the astounding discovery that in higher organisms the coding regions of many genes are interrupted by long stretches of nucleotides that do not code for amino acids found in the protein. The organization of the human β-globin gene is shown in Figure 11-4 as a gene containing three coding regions, called **exons,** interrupted by two noncoding regions, or **introns.** Cases of more than 50 introns scattered through a single gene have been reported. The RNA molecules transcribed from genes containing introns are longer than the messenger RNAs that subsequently produce the protein specified by the gene (Figure 11-5). Cells therefore have a mechanism that removes introns and yields mRNA composed only of coding sequences (exons). This process is called **RNA splicing.**

Although an intron-containing phage gene has been found, in general, bacterial genes do not contain introns; thus splicing is not a part of bacterial messenger RNA processing. The reason for this difference between bacteria and higher cells is not known. Certainly RNA splicing provides many additional options for controlling gene expression. This is particularly apparent with animal viruses, where different patterns of splicing can allow a single region of DNA to produce several different proteins. Splicing is also important for generating antibodies of different types, as discussed later in this chapter.

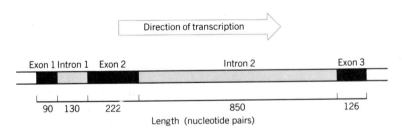

Figure 11-4 Intervening Sequences in the β-Globin Gene of Humans. The exons code for the amino acids in the protein.

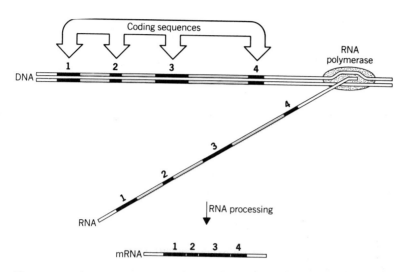

Figure 11-5 Arrangement of Sequences in a Gene from a Higher Organism. The regions of DNA coding for amino acids in the protein product are interspersed with noncoding regions that are removed during formation of mature mRNA. Coding sequences 1 through 4 are part of a single gene. Note that in higher organisms ribosomes do not bind to messenger RNA before it is released from the DNA, as is the case in bacteria.

ENZYMES MADE OF RNA

Earlier chapters stressed that proteins, as enzymes, control cellular chemistry by accelerating specific biochemical reactions. Nucleic acids, on the other hand, were portrayed as repositories for genetic information, repositories whose general function is to direct the synthesis of all cellular proteins. Thus it was a surprise to find that certain RNA molecules can act as enzymes.

One of the first examples of an RNA enzyme was found in ribosomal RNA of a protozoan, a small unicellular animal. Ribosomal RNA molecules are transcribed from DNA as much longer molecules that are later cut and processed. RNA splicing occurs during the processing, and excision of a particular intron was found to occur in test tubes after all proteins had been removed: the intron RNA is self-splicing. After excision, the intron retains its splicing capability and will act on other small RNA molecules. With certain oligonucleo- tides as substrates, the intron even makes oligonucleotides longer. Thus in a sense RNA can synthesize RNA as well as cut it. Such catalytic RNAs are called ribozymes.

The ability of RNA to catalyze chemical reactions has important biological implications. One is that RNA, which is a major compo- nent of ribosomes, may play an active role in the joining of amino acids to form proteins. Another concerns the origin of life. Since RNA molecules can act as enzymes as well as informational molecules, perhaps when life began, RNA functioned without DNA or proteins.

Ribozymes also represent a new way to manipulate and control the biology of living cells—they can be used to cut specific messenger RNA molecules. To be a suitable target for one type of ribozyme, a messenger RNA need only contain the nucleotide sequence GUC. The ribozyme consists of a catalytic region, which is 22 nucleotides long, and short substrate-binding regions that can be any nucleotide sequence. When a ribozyme is made with its substrate-binding re- gions complementary to the RNA target on either side of the target GUC, the ribozyme will bind to the target by complementary base pairing. This will bring the catalytic region into position to cut the target RNA (Figure 11-6). Since an entire ribozyme can be synthe- sized in the laboratory, the nucleotide sequence in each region can be varied at will. Thus a ribozyme can be made that will attack any RNA that contains a GUC (or a small number of other particular nucleotide triplets). The ribozyme can then be delivered to a cell by cloning the DNA template for the ribozyme into a plasmid or virus immediately downstream from a promoter. The intracellular synthe- sis of the ribozyme, and thus destruction of its target, can then be regulated by controlling transcription from the promoter (see Chap- ter 4 for a discussion of transcriptional control). Ribozymes are very specific, and if they are not directed at an RNA essential for cell growth, they cause little harm to the cell. Thus in principle they can be used to protect cells from virus infection or to rid cells of harmful gene products.

ANTIBODIES

Antibodies are proteins that recognize and bind to foreign sub- stances, that is, substances not normally found in our bodies. As

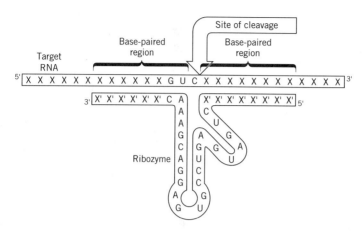

Figure 11-6 An RNA-Cleaving Ribozyme and Its Target. Partial complementarity allows the two RNA molecules shown to base-pair. Any RNA can be the target of the ribozyme shown, as long as the target contains the three-base sequence GUC (or a few other triplets). Cleavage occurs on the 3′ side of the C. This ribozyme is composed of a central catalytic region and two substrate-binding regions. The substrate-binding regions (nucleotides labeled X′ plus AC near the 3′ end of the ribozyme) form base-pairs with complementary regions of the target as indicated and bring the catalytic region into proper position to cleave the target RNA. A ribozyme of the type shown is constructed by first selecting a target region. This is done by scanning the RNA nucleotide sequence of the target gene for a GUC sequence. A particular gene may have several such sequences. Once the target has been chosen, a DNA molecule is chemically synthesized to serve as a template for synthesis of the ribozyme. The DNA usually has a promoter and sites for restriction nucleases at its ends, so it can be easily inserted into a cloning vehicle for delivery into a cell. There RNA polymerase will properly synthesize the ribozyme.

such, antibodies form part of our immune system, the elaborate network of molecules and cells that protects us from many types of disease. When an antibody attaches to a foreign substance, which is called an **antigen,** a number of processes are activated that result in destruction or expulsion of the antigen. Millions of antigens can be recognized by antibodies, each by a different antibody. Each antibody is composed of four protein chains, two identical **heavy chains** and two identical **light chains.** The chains are folded and connected to form a "T" as shown in Figure 11-7. Comparison of amino acid sequences from many different antibodies has revealed several interesting features. First, antibodies can be grouped into classes based on the amino acid sequence and properties of the heavy chains. Second, within a class there are sections of the protein chains that are identical from one antibody to the next. These sections are called **constant regions,** and they determine the behavior of the antibody in our bodies. For example, heavy chain antibodies with one type of constant region circulate in the blood, those with another type attach to the surface of the cell that produced them, and still others bind to specific cells that release histamines. A third point is that each light chain, as well as each heavy chain, has regions of amino acids that are unique to that antibody. These regions are called **variable regions;** they are the parts of the antibody that bind to foreign substances such as viruses and bacteria. Since the shape and structure of a protein are dramatically affected by small changes in the sequence of amino acids, the slight differences in amino acids found in the variable regions result in millions of different antibodies, each able to recognize a particular antigen.

For several decades biologists puzzled over how so many different antibodies could be produced. Were there millions of genes, one for

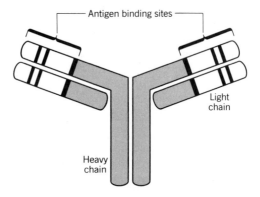

Figure 11-7 Schematic Diagram of an Antibody Molecule. Two heavy chains pair with each other and with two light chains to form the active antibody. The amino acid sequences are divided into constant regions (shaded), variable regions (open), and hypervariable regions (solid). Two antigen binding sites are present, one in the variable region of each arm.

each protein chain of every antibody? Calculations suggested that we might not have enough DNA both to code for all the antibodies and to make up the genes needed to run the chemistry of our cells. DNA sequencing studies now reveal that most of our cells do not have a complete set of antibody genes. Instead, they have bits and pieces that can be combined in a number of different ways, thus producing millions of distinct antibodies from a small amount of genetic information. The rearrangements occur inside special blood cells called **B lymphocytes,** which are responsible for making antibodies.

By comparing the nucleotide sequences in DNA from embryonic cells with those in DNA from antibody-producing cells, it has been possible to develop a general idea about how gene shuffling creates a variety of antibody chains. In the case of light chains (Figure 11-8), the embryo contains several hundred variable region genes (V) widely separated from five short, joining genes (J). DNA breakage and rejoining occur so that one of the V genes is placed next to one of the J genes. RNA polymerase transcribes this region and continues until it also transcribes a constant region gene (C). This long RNA molecule is then spliced to remove the sequence between the V/J region and the C region, producing mature messenger RNA. The messenger RNA is then translated into an antibody light chain. Since any one of perhaps 150 V genes can join to any of 5 J genes, roughly 750 combinations (150 × 5) can occur. Moreover, the joining sites are not precisely located; thus, the actual number of possible combinations is probably closer to 7500.

The same principles apply for heavy chain formation. However, more elements are involved in creating heavy chain diversity (Figure 11-9). In humans there are about 80 V (variable) genes, 50 D (diversity) genes, and 6 J (joining) genes. Thus there are about 24,000 combinations (80 × 50 × 6) that can form. Flexibility in the V/D and the D/J junctions probably adds 100 more ways to combine the genes, so the total number of heavy chain combinations is about 2.4 million (24,000 × 100). The total number of antibody combinations is the product of the light chain and heavy chain combinations, or 18 billion (7500 × 2,400,000). Thus enormous diversity can be produced by about 300 embryonic DNA segments.

As mentioned earlier, there are several types of heavy chain, each with a different constant region that determines how the antibody behaves in the body. Constant region genes are arranged down-

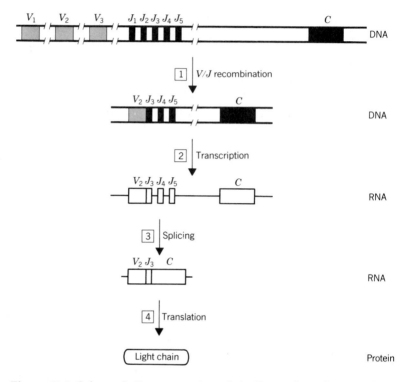

Figure 11-8 Schematic Representation of the Formation of an Antibody Light Chain. **(1)** One of the approximately 150 variable genes *V* recombines with one of the 5 joining genes *J*. In the example V_2 is moved and becomes adjacent to J_3. **(2)** RNA is synthesized from this DNA to produce a primary transcript. **(3)** Splicing occurs to remove all the RNA between J_3 and the constant gene *C*, producing mature messenger RNA. **(4)** This messenger RNA is translated into the antibody light chain. Discontinuities in the DNA indicate large distances between the genes.

stream from the *J* region, and by selective splicing and additional recombination it is possible to put the same *V/D/J* region onto five different constant regions. Consequently, our bodies can contain a number of antibody types that recognize the same foreign substance but perform different functions to attack it.

In summary, we appear to combat foreign substances in the following way. There are millions of B lymphocytes circulating in our blood, and during development and maturation, each of these cells

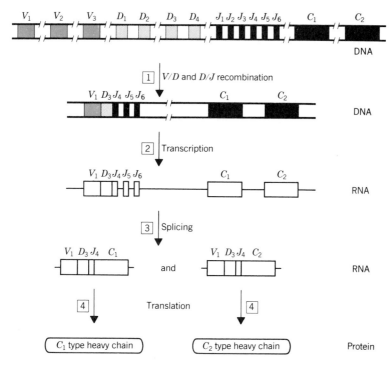

Figure 11-9 Schematic Representation of the Formation of Antibody Heavy Chains. **(1)** One of 80 V regions joins with one of about 50 D regions and one of 6 J regions to form a recombined DNA molecule in a cell called a B lymphocyte. **(2)** A primary transcript is made that contains two different C regions. **(3)** By differential splicing, two types of heavy chain messenger RNA can be made. **(4)** When the messenger RNAs are translated, they produce two types of heavy chain protein. Since the $V_1/D_3/J_4$ regions are the same for both, the two heavy chain proteins will have identical antigen binding sites. Discontinuities in the DNA indicate large distances between the genes.

undergoes a slightly different DNA arrangement. Thus each one produces antibody molecules that are slightly different from those produced by other B lymphocytes. At different times a cell can produce antibodies that differ in their heavy chain constant regions. Thus the antibody can behave differently at different times. The function of one of these classes is to reside on the surface of the B lymphocyte that produced it. There the antibody acts as a sentry, waiting to intercept a specific foreign substance, an antigen. When the antibody on the cell encounters an antigen to which it can bind, a complex is formed between the antigen and the antibody. The complex then triggers that particular lymphocyte to multiply and to produce additional antibody molecules (see Figure 11-10). All antibodies from a particular cell line have identical variable regions, so they all recognize the same antigen. Since the constant regions vary, we end up with antibodies that can function in different ways to rid our bodies of foreign substances. Moreover, a complex antigen, such as a virus, has many different parts called **antigenic determinants.** Each can be recognized by a different B lymphocyte; consequently, the proliferating population of B lymphocytes will produce a population of antibodies capable of recognizing many different aspects of the virus. This increases our chances of fighting off infection.

The type of immune response just described is called **polyclonal** because many different clones of cells arise. Methods have been developed to grow clones from single B cells, clones that produce large quantities of a single type of antibody. Such antibodies, called **monoclonal** antibodies, are rapidly becoming valuable research and clinical tools because they are so specific. Efficient laboratory production of monoclonal antibodies involves the fusion of two different types of cell. First, an animal is immunized with the antigen against which one wishes to produce antibodies. This is generally achieved by means of several injections of the antigen over a period of weeks or months. A collection of B cells is then obtained, usually by dissecting the animal's spleen. The B cells are mixed with cultured **myeloma cells** under conditions in which the two cell types fuse to form what is called a **hybridoma.** Myeloma cells are malignant (cancerous) antibody-producing cells. When fused to a B cell, a myeloma cell stimulates the B cell to secrete large amounts of the antibody that the B cell had been programmed to produce. The myeloma cell also immortalizes the B cell (cultured B cells normally die within a few days after removal from the body).

Of course the fusion mixture contains many uninteresting cell types and fusion products. To allow only the hybridomas to grow in culture, genetic defects have been introduced into the myeloma cells. The strategy is similar to the use of antibiotic resistance in

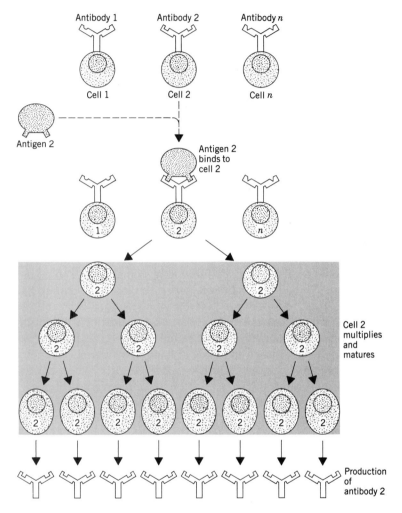

Figure 11-10 Clonal Selection for Antibody Production. Each antibody-producing B lymphocyte (cells 1, 2, . . . , n) produces just one type of antibody, which is placed on the cell surface. Each B lymphocyte recognizes a different antigen. When an antigen comes in contact with a B lymphocyte in the proper way and is recognized by that cell, the cell is stimulated to divide and produce antibodies. Thus a large number of cells producing antibodies that recognize a specific antigen is generated by a single contact.

213

gene cloning to obtain a particular desired bacterial colony. Once the culture has been enriched for hybridomas, single cells are cultured separately and allowed to multiply. The resulting cultures are then tested for the production of the desired antibody. In this procedure, myeloma variants that have stopped producing their own antibodies are used; thus, the resulting hybridomas produce only the antibody made by the B cell.

Specific antibody–protein interactions play the same role in the study of protein biology that nucleic acid hybridization plays in the analysis of genes. Antibodies can be used to separate specific proteins from other **macromolecules,** to determine the location of a protein in the cell, to quantify various proteins in body fluids, and even to clone genes. Monoclonal antibodies greatly enhance the precision of analysis because the antibody preparation recognizes only a single determinant on the protein. Moreover, once cultured, the cells producing the monoclonal antibodies always produce the same antibody; this consistency is not found when animals are used as antibody sources, since their antibody specificities often change over time.

In clinical medicine the potential use of monoclonal antibodies ranges from diagnosis of diseases to therapy. Their exquisite specificity makes antibodies well-suited for identifying cells of specific types. For example, antibodies that recognize tumor cells could be used to locate tumors and then evaluate the effectiveness of surgery or chemotherapy. Small cytotoxic molecules could even be attached to the antibodies; when the antibodies are injected into cancer victims, the antibodies would attach to the cancer cells and kill them. Infectious diseases could also be fought with monoclonal antibodies by injection either before or soon after exposure to dangerous viruses and bacteria. Indeed, this form of passive immunization is one of the approaches being taken to try to combat AIDS.

TRANSPOSITION

Gene cloning technologies allow us to insert small, discrete fragments of DNA into specific places in other DNA molecules. Nature also has this ability. Scattered among the genes of living cells are small, discrete sequences of nucleotides that can hop from one region of DNA to another or from one DNA molecule to another. These discrete nucleotide sequences are called **transposons,** and the process in which a transposon moves to another location in DNA is called **transposition.** In some types of transposition a duplication of the transposon occurs, as in Figure 11-11, while in other types the transposon excises and moves to a new location.

Transposition can have several consequences, depending on where the transposon inserts. For example, when a transposon inserts into the coding region of a gene, the information in the gene is interrupted (gene Y, Figure 11-11), and the gene may no longer produce a functional protein. Cases have also been observed in which transposition activates a gene by insertion near the gene.

Transposons have been found in a wide variety of organisms, and it is likely that all organisms contain them. Transposons from different organisms appear to share three common features. First, transposons are always discrete sections of DNA; the junctions between a transposon and the DNA in which it is inserted are precisely defined. Second, transposons usually contain nucleotide sequences encoding one or more protein products required for movement of the transposon from one site to another. Thus transposons contain

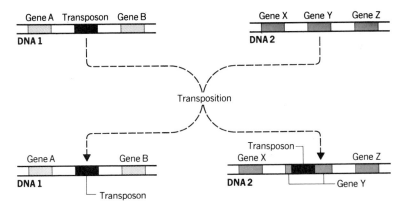

Figure 11-11 Transposition. In DNA 1, a specific region called a transposon (solid region) duplicates itself and inserts a copy into DNA 2. In this example the transposon inserts in the middle of gene Y, splitting the gene into two parts.

genes responsible for their own movement. Third, each end of a transposon contains nucleotide sequences that probably serve as recognition sites for factors involved in movement of the transposon. These sequences are often repeats of each other; in most cases the repeats are inverted. A general scheme of transposon structure is shown in Figure 11-12.

Although the molecular details of transposition vary from one type of transposon to another, a description of one called Tn3 is sufficient to provide an appreciation for the process. Tn3 is found in bacterial cells, and it contains three genes, *A*, *R*, and *bla* (Figure 11-13). In addition, Tn3 has a stretch of 38 base pairs at its left end that is repeated at its right end in an inverted orientation. The *bla* gene encodes a protein that destroys ampicillin (penicillin); thus any cell containing Tn3 is resistant to ampicillin. This feature helps biologists determine whether a cell contains Tn3. The *A* gene encodes

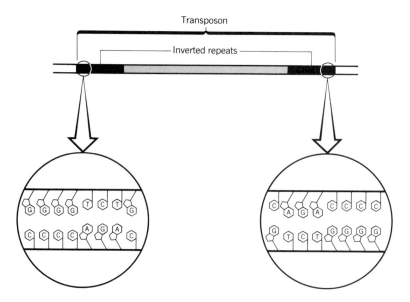

Figure 11-12 General Structure of a Transposon. Transposons contain repeated nucleotide sequences at each end. The repeated sequences are generally in an inverted orientation. Genes involved in transposon movement lie between or within the repeats.

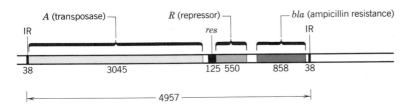

Figure 11-13 Arrangement of Genes in Tn3. This transposon contains three genes, *A, R,* and *bla,* located between the 38-base-pair inverted repeats (IR). The repressor protein binds to Tn3 DNA at region *res.* Numbers indicate nucleotide pairs in the specified regions.

a **transposase,** a protein responsible for Tn3 movement. If small regions of the *A* gene are experimentally removed, Tn3 can no longer move. The *R* gene codes for a repressor that binds to the *A* gene and prevents it from making transposase. Thus the repressor keeps transposition from occurring very often.

It is likely that transposition of Tn3 occurs as a two-step process. In the first step (Figure 11-14) the DNA molecule containing Tn3 (the donor DNA) somehow binds to a DNA molecule lacking Tn3 (the recipient DNA). This process is mediated by the transposase protein. Then the Tn3 sequence is duplicated to produce a structure called a **cointegrate.** The cointegrate contains two copies of Tn3, one at each junction between the donor and recipient DNAs. After the cointegrate has formed, transposase dissociates from the DNA.

In the second step the cointegrate is split into two circles, and both DNA molecules, donor and recipient, end up with a copy of Tn3. The product of the *R* gene, the repressor, plays a key role in this process (Figure 11-15). It binds to the cointegrate DNA and aligns the two copies of Tn3. Breaks then occur in the two copies of Tn3 at the *res* sites. One double-stranded DNA crosses over the other, and the broken ends are joined so that a part of one Tn3 is linked to the remainder of the other. This process of DNA strand exchange is called **recombination.** In this case, recombination also creates two DNA circles, each with a copy of Tn3. Thus the repressor plays two roles in transposition of Tn3: first, it keeps the frequency of transposition low by repressing the *A* gene, and second, it is required to align, break, and rejoin cointegrate DNA. The repressor

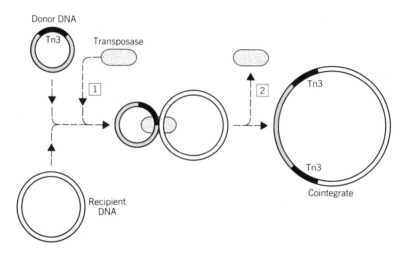

Figure 11-14 Scheme for Formation of Cointegrates by Tn3. (1) Transposase mediates the joining of a DNA molecule containing Tn3 (donor DNA) with a DNA lacking Tn3 (recipient DNA). **(2)** Tn3 is replicated (duplicated by DNA polymerase), and in the process the donor and recipient DNAs are joined to form a larger circle (cointegrate). The transposase leaves the DNA and is presumably free to initiate a new round of transposition.

also represses the *R* gene—it represses its own production. Transposition is a very tightly regulated process.

Transposons are among the more exciting tools for genetic research, since they offer a way to insert genes into animals. A fruit fly experiment provides an example. A transposon from a fruit fly was first cloned into a bacterial plasmid. Then transposon-containing plasmid DNA was isolated from the bacteria, and a fly eye color gene was inserted into the transposon. The engineered transposon, still on the plasmid, was returned to the bacterium, permitting investigators to obtain large quantities of the engineered transposon. Purified plasmids containing the transposon were then injected into fruit fly embryos. There the transposon, along with the new eye color gene it carried, transposed into a fruit fly chromosome. When the fly developed into an adult, it had the eye color dictated by the gene that had been inserted into the transposon.

Transposons may also be important for their ability to inactivate genes. Some genetic diseases may be due to the inactivation by

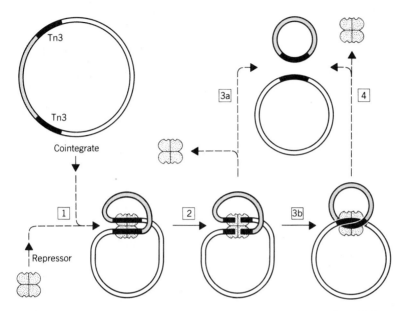

Figure 11-15 Recombination Between Two Tn3 Transposons in a Cointegrate. (1) The repressor protein binds to the cointegrate and probably uses the natural twists in the DNA to align the two copies of Tn3 (solid). **(2)** A break occurs within the *res* site (see Figure 11-13) of each Tn3. **(3)** One DNA crosses over the other, and the breaks are sealed. This produces two rings: the donor (shaded) and the recipient (open), each containing a copy of Tn3. Recombination has two possible outcomes: two separate rings are produced directly **(3a)** or two interlocked rings arise **(3b)**. In both cases the repressor dissociates from the rings **(3a, 4)**. **(4)** An enzyme called a topoisomerase will separate the interlocked rings (see Figure 2-7).

transposons of crucial genes, and if we had a way to remove the transposon, we might be able to reactivate the gene and thus cure the disease.

CROWN GALL TUMORS AND THE Ti PLASMID

Plants can develop tumors, and an example arising from the crown gall disease is shown in Figure 11-16. Crown gall tumors are caused by a large bacterial plasmid named Ti, for tumor inducing, whose

Figure 11-16 Crown Gall Tumor. A turnip root was inoculated a single time with a virulent strain of the bacterium *Agrobacterium tumefaciens*. Tumor cells initially developed at the site of inoculation and then spread. The bacteria can be removed from the tumor cells, and the tumor cells can be cultured on agar lacking growth hormones (normal cells do not grow under these conditions). When inoculated into a healthy plant, the cultured tumor cells will generate a new tumor. (Photo courtesy of Dr. C. I. Kado, University of California, Davis.)

normal host is the bacterium *Agrobacterium tumefaciens.* The tumors result from an intimate relationship among the plant, the bacterium, and the plasmid. For tumor formation to occur, the living bacterium must gain access to the plant through a wound; then part of the Ti plasmid is transferred into the plant cells. The plasmid contains three genes encoding proteins required for the production of plant growth hormones; when these hormones are produced, the plant cells begin multiplying uncontrollably. The plasmid also contains genes (*nos*) that encode proteins required for the production of compounds called opines. Other genes (*noc*) on the plasmid specifically break down the opines into compounds that can be used by the infecting bacterium as a nutrient source. Thus the bacterium benefits from the uncontrolled cell growth and production of opines. The plant cells become genetically altered by the uptake of the plasmid DNA, and indeed tumor cells can be cultured indefinitely in the absence of the infecting bacterium or added growth hormones.

The Ti plasmid–*Agrobacterium tumefaciens* transformation system has recently been used to genetically manipulate plants. Up to 50,000 base pairs of foreign DNA can be inserted into the plasmid and subsequently incorporated into the plant DNA. This method has produced engineered foods such as tomatoes that maintain their firmness while ripening.

PERSPECTIVE

DNA must have a stable structure, since this substance is the vehicle through which the characteristics of a species are accurately passed from one generation to the next. But DNA is not static, either in information content or in three-dimensional shape. As pointed out above, gene duplications probably occur, and over time mutations arise that lead to differences in nucleotide sequence between duplicated genes. These differences could allow the protein products to perform different functions. Sequence relatedness among exons of different genes now suggests that duplicate copies of genes may be shuffled and joined to form new, larger genes that can provide cells with new proteins having multiple functions. Whether transposons contribute to this process is unknown, but it is clear that they hop around in the DNA, constantly creating the potential for changing gene structure. Antibody gene rearrangements show that even within the lifetime of an individual, gene rearrangements can be quite common.

Viruses and plasmids are responsible for another aspect of DNA dynamics, the movement of DNA from one cell to another. With viruses, gene transfer occurs when host DNA is mistakenly packaged inside virus particles and then carried to a new host upon infection by the virus. In bacteria some plasmids integrate (insert) into the host DNA. The plasmids then either mobilize the host DNA for direct transfer or improperly excise, with the result that a piece of host DNA remains a part of the plasmid. Release of the plasmid into the environment would allow it to infect a new host. The transfer of the Ti plasmid involves not only plasmid genes but also an intimate relationship between the bacterial and plant hosts. When retroviruses are involved, as is often the case with gene transfer in animals, the transfer includes a step in which the genes are in the form of RNA rather than DNA. There are even transposons that are very similar in biology to a degenerate retrovirus, one that no longer escapes from the cell. Because of their potential for human genetic engineering and their roles in AIDS and human cancer, retroviruses currently command a great deal of public attention.

Questions for Discussion

---

1. Why are people unlikely to show an immunological response to poison oak or poison ivy the first time they touch these plants?
2. If you were to compare nucleotide sequences between monkey and man, would you expect to find more differences in genes or pseudogenes?
3. Most human genes are not composed of a continuous stretch of DNA; instead they contain alternating stretches of coding DNA (exons) and noncoding regions (introns). Both the exons and the introns are transcribed into RNA, but the introns are removed via RNA splicing. Sometimes the amount of DNA devoted to introns far exceeds that in exons. What purposes

might the introns have that would justify the biochemical cost of synthesizing them?

4. RNA molecules that will specifically cleave other RNA molecules are called ribozymes. One type of ribozyme cleaves next to the trinucleotide GUC. How might synthetic ribozymes be useful in fighting virus infections? What information would you need to have to design an effective ribozyme?

5. For a ribozyme to be effective against a virus, it must be present inside the cell. How might you generate a high concentration of a specific ribozyme inside a cell?

DNA concentration μgs/ml

= dilution × Absorbance × 50
  factor   at 260   constant.

Log 10 of Mr.

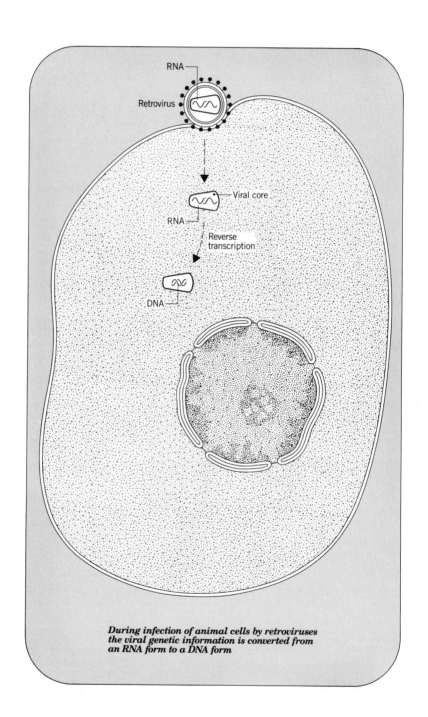

During infection of animal cells by retroviruses the viral genetic information is converted from an RNA form to a DNA form

C H A P T E R   T W E L V E
RETROVIRUSES
AIDS and Oncogenes

Overview _____

Retroviruses comprise a group of related viruses in which the viral genetic information is converted from an RNA form to DNA during infection of animal cells. As DNA the viral information becomes inserted into host chromosomes. Then viral DNA directs the synthesis of viral RNA and proteins, which spontaneously assemble to form progeny virus. Viruses of this type have been detected in many animal species. One retrovirus causes AIDS, killing cells normally involved in the human immune response. Others are associated with cancer, a general term used for a variety of diseases involving uncontrolled cell proliferation. In addition to their role in disease, retroviruses serve as cloning vehicles and provide a source for reverse transcriptase, an enzyme used in genetic engineering to convert RNA to DNA.

INTRODUCTION

During the life cycle of retroviruses there is a stage at which genetic information flows from RNA to DNA, a direction opposite to that found in most genetic systems. This reverse, or retrograde, flow of information is the basis for the name "retrovirus." Once in the chemical form of DNA, the viral genetic information is inserted into the DNA of the infected cell. There it can remain for long periods of time as a molecular parasite. In principle, this insertion or integration process is similar to the formation of bacterial lysogens by bacteriophage lambda (see Chapter 6 and Figure 6-8), although insertion does not occur at a specific site as it does with lambda.

Retroviruses merit a chapter to themselves because of their widespread importance. Most of their notoriety has come from diseases they cause in humans. One example is a deadly form of leukemia caused by a virus called HTLV-I, and another is the acquired immune deficiency syndrome (AIDS) caused by the human immunodeficiency virus type 1 (HIV-1). AIDS is a lethal disease in which the immune system is debilitated to such a degree that it cannot protect the victim from microorganisms that are normally innocuous. Other retroviruses are called RNA tumor viruses because they carry **oncogenes,** genes that cause malignant cell growth.

Less obvious, but perhaps equally important in the long run, are the roles played by retroviruses in gene cloning and gene therapy. For example, retroviruses are the source of reverse transcriptase, an enzyme tool that makes it possible to synthesize DNA from RNA molecules (since nucleotide sequences are much easier to determine and manipulate as DNA than as RNA, reverse transcriptase has been a key for the study of RNA as repositories for genetic information and as biological catalysts). Retroviruses also serve as cloning vectors for placing foreign genes into human cells; they give us the potential to permanently change genes in humans.

Our understanding of retroviruses and their interactions with their animal hosts is far from complete, and many paradoxes remain to be resolved. But firm statements can be made about the molecular anatomy of the viruses and certain aspects of their life cycle. Some of these details are presented below to provide the understanding necessary for readers to make future decisions concerning these exceptionally important agents.

RETROVIRUS STRUCTURE

Retrovirus particles are spherical structures (Figure 12-1*a*) containing two RNA molecules, each about 10,000 nucleotides long, and at least nine different types of protein, some of which are named in Figure 12-2. The RNA is packaged by viral proteins into a structure called a core. The proteins of the core, which are encoded by the viral gene called *gag*, are among the major constituents of the virus. The viral core also contains the reverse transcriptase needed to synthesize DNA from RNA, a site-specific protease, and a protein called **integrase.** Integrase participates in inserting the viral DNA into the host DNA. As indicated in Figure 12-1*b*, the core is surrounded by an **envelope** composed of a double layer of **phospholipids** (fatlike molecules) derived from the surface membrane of the host cell in which the virus was formed. Within the envelope are embedded proteins of several types that are encoded by the viral gene called *env.* Some of the viral proteins in the envelope specifically interact with cellular proteins called receptors. These receptors are located on the surface of animal cells, and the interaction of the viral envelope proteins with cellular receptors enables the virus to infect specific cell types.

Retroviruses have a common genetic organization (Figure 12-3*a*). At the ends of the viral RNA are blocks of repeated nucleotides called R. Internal to R at the 5' end of the RNA is a block called U5; at the 3' end is another block called U3. U5 and U3 are duplicated when the RNA is converted into the DNA form of the viral genome (see below), and the resulting DNA copy has 5' U3-R-U5 3' regions at each end. These sequences, called **long terminal repeats (LTRs),** are important because they contain the viral promoter from which viral RNA is synthesized. The outer ends of the LTRs contain nucleotide sequences crucial for insertion of the viral DNA into the host chromosome.

The *gag* gene is nearest to the 5' end of the RNA, and it encodes a long **polyprotein** from which several proteins are subsequently cleaved. These proteins are part of the viral core, and they probably serve to wrap and protect the viral RNA. On the 3' side of *gag*, and sometimes overlapping it slightly, is the gene called *pol.* This gene also encodes a long polyprotein, which is actually synthesized

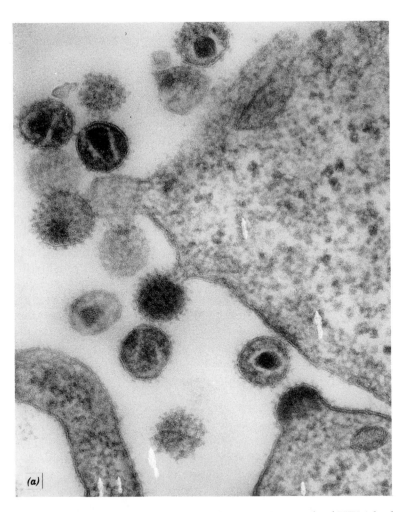

Figure 12-1 Retrovirus Structure. (a) Electron micrograph of HIV-1 budding from a human cell; magnification is 120,000 times. Infected cells were cut into thin sections prior to microscopic examination. The particles reveal different features of viral structure because they were not all cut at the same place. [Photomicrograph courtesy of Dr. H. Gelderblom, Robert Koch Institute, Berlin, reprinted from *Arch. Virol.* 100:255–266 (1988), with permission.] **(b)** Diagrammatic representation of a retrovirus in cross section. The outer shell or envelope is composed of a double layer of lipid (solid) and protein. In HIV the major envelope glycoprotein has knobs that are called gp120, and this protein is anchored to the envelope by a transmembrane protein called gp41. The Env proteins exist in the membrane as **tetramers.**

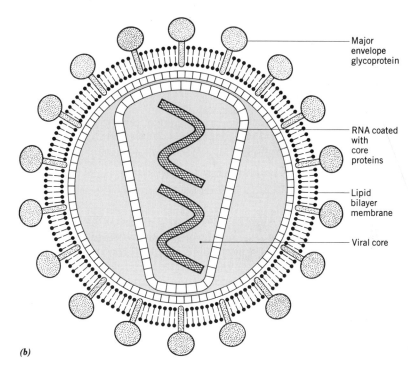

(b)

Figure 12-1 (*Continued*). Inside the envelope is a core consisting of several types of protein from the *gag* gene that package and protect two RNA molecules. The respective positions of the two RNA molecules and their mutual orientation are not known. Also present in the core are the viral reverse transcriptase and integrase. (Adapted from W. Hazeltine and F. Wong-Staal, *Scientific American*, October 1988.)

as a very long *gag-pol* polyprotein. In the case of HIV-1, the *pol* gene product is subsequently cut into three proteins: a specific protease responsible for cutting the polyproteins, reverse transcriptase, and integrase. The third gene, *env*, encodes a polyprotein which, when cleaved, produces two viral proteins that eventually become a part of the virus envelope. In the case of HIV-1 these two proteins are called gp120 and gp41 (gp stands for glycoprotein, i.e., sugar-containing protein; the numbers refer to the sizes of the proteins).

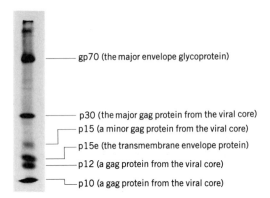

gp70 (the major envelope glycoprotein)

p30 (the major gag protein from the viral core)
p15 (a minor gag protein from the viral core)
p15e (the transmembrane envelope protein)
p12 (a gag protein from the viral core)
p10 (a gag protein from the viral core)

Figure 12-2 Display of Retroviral Proteins. Particles of the FrMCF virus were isolated from mouse cells, and the preparation was treated with detergent to disrupt the chemical interactions that hold the viral components together. The viral proteins were then separated by gel electrophoresis using methods similar to those described in Figure 7-4. An electric current forced the proteins to move through a gel of acrylamide. Under the conditions used, smaller proteins moved faster than larger ones. In the figure, the direction of migration was from top to bottom. This particular preparation was radioactively labeled, and after electrophoresis the gel was used to expose X-ray film. Dark spots on the film indicate locations of proteins. The proteins have been given names based on their size; thus analogous proteins from different viruses often have different names. The letter "g" indicates that sugar groups are attached to the protein. Some proteins (e.g., reverse transcriptase, integrase, viral protease) are present at such low concentration that they do not show up with this type of analysis. (Photo courtesy of William Honnen and Abraham Pinter, Public Health Research Institute.)

The overall genetic organization just described is common to retroviruses, but variations on the theme occur. Viruses of the type that cause AIDS have several additional elements that control viral gene expression, and viruses that cause tumors tend to have parts of their genome replaced with animal genes. These special cases are described in more detail in later sections.

RETROVIRUS LIFE CYCLE

Our ability to grow animal cells in culture and infect them with viruses has made it possible to study the life cycles of many

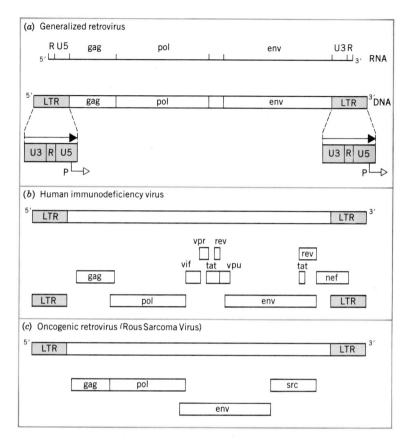

Figure 12-3 Genetic Organization of Retroviruses. (a) Although each retroviral species has a distinct nucleotide sequence, retroviruses share a common genetic structure. The region called R is a repeat found at both ends of the RNA. Regions called U5 and U3 occur only at 5′ and 3′ ends, respectively, of the viral RNA, but they are both present at each end of the DNA. The U3-R-U5 arrangement creates two long, terminally repeated sequences (LTRs), and the DNA is longer than the RNA by the number of nucleotides in U5 and U3. The positions labeled **P** are strong promoters, and the direction of transcription is indicated by arrows. The three genes *gag, pol,* and *env* are common to all retroviruses. The 5′ and 3′ notations on the DNA are for the coding strand. **(b)** The AIDS virus (HIV-1) has the standard retrovirus organization plus additional genes called *vif, vpr, vpu, tat, rev,* and *nef.* Gene overlapping, as indicated schematically here, is not uncommon. The *tat* and *rev* genes are split, and their expression requires RNA splicing. **(c)** The oncogenic Rous sarcoma virus contains a gene called *src* on the 3′ side of *env.* The product of *src* is thought to perturb the control of cell division, leading to formation of a sarcoma, a type of tumor.

231

animal viruses including those that cause polio, influenza, and AIDS. In one method of culturing cells, a piece of tissue is surgically removed from an animal. The tissue piece is then placed in a plastic petri dish containing a nutrient medium and treated with enzymes to detach the cells from each other. Some cell types grow and divide in suspension, while others stick to the bottom of the dish. In the latter case, the cells stop growing and dividing when they come into contact with each other. Normally they form a monolayer on the bottom of the dish. In either case, dilution and transfer of the cells to another petri dish leads to renewed growth and division. In this way some types of animal cell can be perpetuated for many generations. Addition of virus particles to the cell culture results in infection and eventually in release of newly made virus into the culture medium. Some viruses cause infected cells to die. HIV-1, which infects cells growing in suspension, falls into this category. Cells infected with HIV-1 may also fuse with other infected cells to form giant cells containing many nuclei.

The first stage of retroviral infection is the entry of the virus into an animal cell (Figure 12-4*a*). This appears to occur by a fusion of the viral envelope with the cell membrane, leading to the movement of the viral core into the cytoplasm of the cell (Figure 12-4*b*). Specific proteins on the cell surface serve as receptors for the virus and allow penetration to occur. Consequently, only certain types of cell are infected, those containing the receptor protein. For example, HIV-1, the AIDS virus, preferentially infects specific cells of the human immune system by interacting with a particular receptor protein, called CD4, which is present on the surfaces of these cells.

Soon after entry into the cell, the viral reverse transcriptase (a DNA polymerase) makes a double-stranded DNA copy of the information in the viral RNA (Figure 12-4*c*). The viral DNA then inserts into the host chromosome by a process called **integration** (Figure 12-4*d*). Both reverse transcription and integration are described in more detail below. Messenger RNA is later synthesized from the integrated virus (Figure 12-4*e*). Some of these RNA molecules serve as the genetic information in new virus particles. Others are translated into viral proteins. For some genes RNA splicing occurs. Viral proteins and full-length viral RNAs then spontaneously assemble

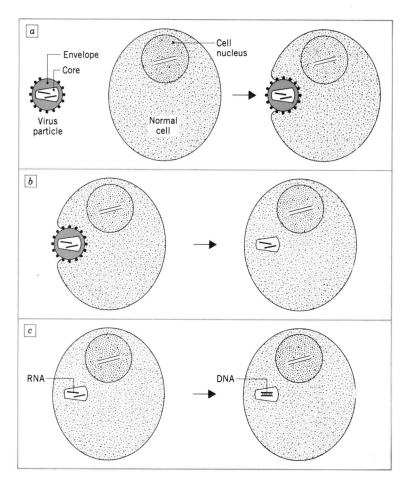

Figure 12-4 Retrovirus Life Cycle. (a) Attachment of virus to an animal cell. Specific proteins on the virus surface are thought to interact with specific receptors on the cell surface, leading to binding of the virus to the cell. **(b)** Viral penetration of an animal cell. The viral membrane fuses to the animal cell membrane, releasing the viral core into the cytoplasm. **(c)** Reverse transcription. Viral RNA is converted into DNA by reverse transcriptase. **(d)** Viral integration. Viral DNA, probably still packaged in the core, migrates to the nucleus of the cell and inserts into an animal chromosome. **(e)** Viral gene expression. RNA is transcribed from the integrated form of the virus. Transcription begins at the strong promoter in the left LTR and continues into the right LTR (see Figure 12-3a). This RNA is used to translate viral proteins. For some genes an additional RNA splicing step is involved.

(Continued)

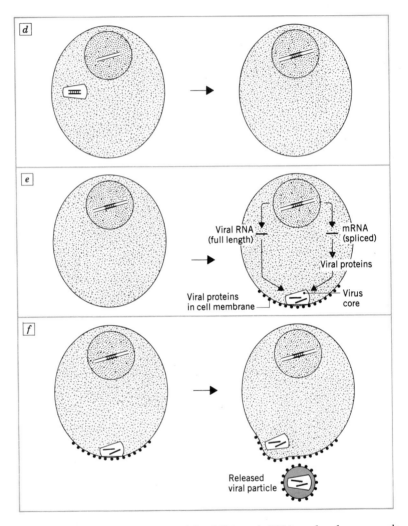

Figure 12-4 (*Continued*). Some of the full-length RNA molecules are packaged into cores by viral proteins. Viral proteins from the envelope gene insert into the cell membrane of the host cell. (**f**) Virus release. Viral cores containing RNA form near the cell membrane. They then appear to be surrounded by the cell membrane, which at that point contains viral surface proteins. In one sense, the viral cores bud out of the cell, covered by cell membrane. Some of these budding steps can be seen in Figure 12-1*a*.

into viral cores. The envelope proteins appear to be incorporated into the **membrane** of the cell, and as the viral cores leave the cell, they are wrapped by a protective coat derived from the cell membrane (Figure 12-4*f*; also see viruses budding from cells in Figure 12-1*a*). With most retroviruses, the infective process does not kill the cell; HIV-1 is an exception.

Reverse transcription occurs inside the viral core, after the core has entered the host cell. Much of the research effort on reverse transcription has focused on two problems:

1. How do the U5 and U3 regions, which are at opposite ends of the RNA, come to lie next to each other in the LTRs of the DNA?
2. How is DNA synthesis primed? (DNA polymerases require a primer to begin synthesis; see Chapter 3.)

One possible scheme is sketched in Figure 12-5. As pointed out in the preceding section, the viral RNA contains a repeated sequence R at each end, as well as subterminal regions called U5 and U3 (see Figure 12-3*a*). There is a region just beyond the 3′ end of U5 that is complementary to a specific transfer RNA (tRNA), and it appears that formation of a tRNA–retroviral RNA hybrid generates the primer necessary for reverse transcriptase to begin DNA synthesis (see Figure 12-5*a*). Reverse transcription then proceeds toward the 5′ end of the viral RNA, stopping when the end is reached. The reverse transcriptase protein contains an activity called **ribonuclease H** (RNase H) that breaks down RNA hybridized to DNA. Ribonuclease H destroys the viral RNA once it has been used as a template, leaving only the DNA copy of the viral genetic information. Removal of the R region of the RNA frees the new DNA strand to hybridize with the R region of the second RNA molecule in the viral core. This may explain why there are two copies of RNA in the viral core. Strand transfer (Figure 12-5*d*) creates a new primer that can be used to continue synthesis to the end of the second RNA.

Next, the second strand of the DNA must be synthesized. This synthesis is thought to be primed by a short oligonucleotide that binds just outside the U3 region. DNA synthesis from this primer results in a short piece of DNA that includes the site where the

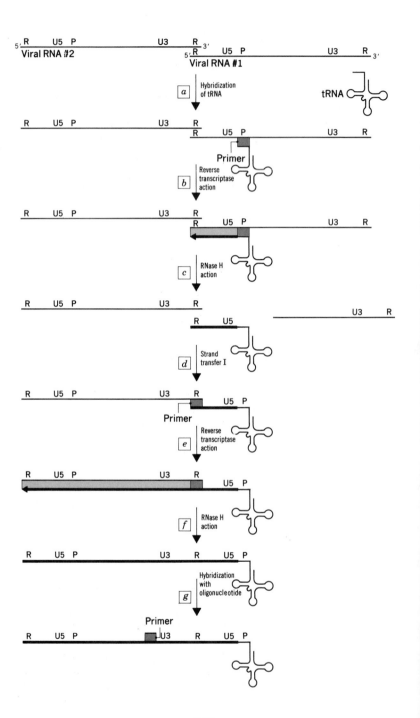

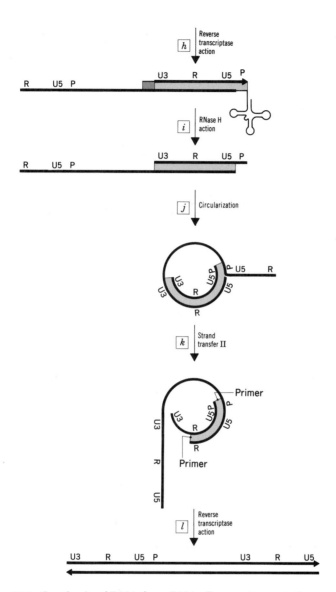

Figure 12-5. Synthesis of DNA from RNA. Reverse transcription converts the genetic information in two viral RNA molecules (thin lines) into a single molecule of DNA (heavy lines). In the process, regions labeled U5 and U3 are duplicated, and with region R they form the direct long terminal repeats called LTRs. One way this might happen is shown. **(a)** The viral RNA and a specific transfer RNA hybridize to create a primer on the RNA close to

(Continued)

237

tRNA binds to RNA, a location called the primer binding site. Circularization could lead to a second strand transfer and the generation of two more primers. DNA synthesis from these primers would then complete the process.

After the viral genetic information has been converted to a DNA form, it integrates (inserts) into the host DNA by a process that has features resembling bacterial transposition and the integration of lysogenic bacteriophages. Mutations, either at the end of an LTR or within the integrase gene, block integration; thus these regions are crucial for integration. It is thought that the ends of the LTRs serve as binding sites for the integrase protein; after insertion they are located at the ends of the integrated **provirus** (Figure 12-6). Integration occurs at a wide variety of places in the host DNA, and at each integration site a few base pairs of host DNA are duplicated. Thus there is a direct repeat of host DNA at the junctions of host and viral DNA. In its integrated form the virus can remain dormant for many years.

AIDS

Almost everyone diagnosed with AIDS possesses antibodies directed against HIV-1. This, plus the appearance of AIDS following

---

Figure 12-5 (*Continued*). U5 at the site labeled **P.** **(b)** Reverse transcriptase then synthesizes a short DNA strand from this primer. **(c)** RNase H action degrades regions of RNA hybridized to DNA. **(d)** Strand transfer occurs by hybridization of **R** regions of RNA and DNA. **(e)** Reverse transcription starts again, this time from the primer created by the 3' end of the DNA. This generates one LTR (U3-R-U5). **(f)** The RNase H activity of the reverse transcriptase degrades the hybridized RNA, leaving only the DNA copy plus the tRNA. **(g)** An oligonucleotide hybridizes to a region adjacent to U3, forming a primer. **(h)** Reverse transcriptase action synthesizes DNA including the primer binding site **(P).** **(i)** RNase H action again degrades RNA hybridized to DNA, releasing the tRNA primer. **(j)** The DNA circularizes through hybridization of the regions labeled **P.** **(k)** A second strand transfer occurs to generate two primers. **(l)** Reverse transcription synthesizes the remainder of each strand.

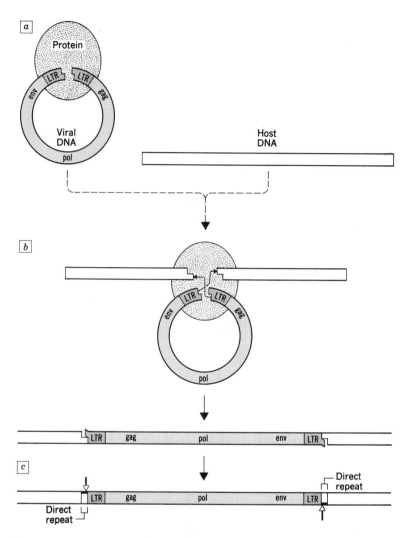

Figure 12-6 Integration of Retroviral DNA. (a) The cell nucleus contains retroviral DNA in a form that can react with chromosomal DNA. Although both circular and linear viral forms of DNA are present, it is the linear DNA that integrates. Following reverse transcription, integrase trims two base pairs from each 3′ end of the viral DNA. **(b)** A protein-mediated complex forms between viral and host DNA, opening the host DNA with a staggered cut (there are five to nine overhanging bases, depending on the virus). **(c)** The viral DNA strands are ligated to the host DNA. The two bases overhanging at each end of the virus are removed, and the single-stranded regions are used as templates for DNA polymerase to fill the gaps (arrows). Thus the inserted virus is bounded by direct repeats and has lost two base pairs from each end.

239

transfusions with HIV-contaminated blood, first suggested that HIV-1 is responsible for the disease. The presence of the antibodies now serves as a clinical test for people harboring the virus. Many "seropositive" people show no symptoms of disease, but it is anticipated that all, or almost all, will eventually exhibit a frank case of AIDS. During the early stages of HIV-1 infection there is often abundant replication of the virus, producing free virus in several parts of the body. This first wave of viral attack is often accompanied by fevers, rashes, and flu-like symptoms. But within a few weeks, the virus levels and symptoms drop, apparently in response to activation of the host immune system. Several years later, the virus replication eventually overwhelms the host defense systems and the final stages of disease set in.

A major cellular target of HIV-1 is a type of blood cell called a helper T cell. This type of cell normally participates in the immunological response of animals against bacteria, viruses, and other microorganisms. Helper T cells called CD4 lymphocytes have on their surface the protein called CD4 or T4, which is also the receptor for HIV-1. Thus they are susceptible to infection and killing by HIV-1. AIDS patients have an abnormally low number of these cells, presumably because of the HIV-1 infection. The loss of these cells leaves the patient susceptible to infection by variety of organisms that normally do not harm healthy people.

HIV-1 has genetic and biochemical properties similar to those of other retroviruses but with additional features. Examination of the nucleotide sequence of the viral genome reveals six regions (*vif, vpr, vpu, tat, rev,* and *nef:* see Figure 12-3*b*) that are not found in most other retroviruses. These regions all encode proteins, several of which cause specific antibody production by AIDS patients. The Tat protein causes a 1000- to 2000-fold increase in transcription originating from the promoter located in the left LTR region. Since transcription of RNA for all the viral genes begins at this promoter, Tat increases expression of all viral genes. Viruses containing a mutation in *tat* show little infectivity. The Rev protein is also a positive regulator. It selectively favors production of proteins that will eventually become a part of the virus. Nef appears to be a negative regulator, inhibiting transcription of full-length viral RNA. The interplay of these regulators probably enhances the ability of the virus to grow rapidly or slowly at different stages of the infection cycle. The product of the *vif* gene appears to be involved in the ability of the virus to infect cells rather than in the ability to replicate. At present little can be said about *vpr* and *vpu*.

Although it is not completely clear why T4 lymphocytes are selectively killed by HIV-1, the viral envelope protein appears to be a key player through its interaction with the high concentration of CD4 receptors present on these cells. According to one scenario, the viral envelope protein accumulates on the surface of infected cells during infection, binds to receptors on nearby uninfected cells, and effectively tears holes in their membranes. This structural damage would cause a cell to leak to death. In another type of cell death, mediated by the envelope protein and the CD4 cellular receptor, an infected cell fuses with nearby cells. A single infected cell can fuse with as many as 500 uninfected cells to form a giant cellular aggregate that eventually dies. A third process of killing is carried out by the immune system itself. Infected cells normally shed large amounts of the envelope protein (gp120), and in addition to stimulating antibody production, the envelope protein probably binds to CD4 receptors on uninfected cells. Consequently, host antibodies against gp120 would bind to these cells. Then other components of the immune system would recognize these uninfected, antibody-covered cells and kill them. Thus only one in a thousand T4 lymphocytes may actually be infected, but the viral envelope protein causes the whole population to be devastated.

CANCER

Cancer is a broad category of disease that involves uncontrolled cell proliferation. As our bodies develop, cells receive signals that cause most of them to stop dividing. Those that do continue to divide are balanced by an equal number that die. Thus as adults we maintain a roughly constant size and shape. It turns out that the products of a small number of special genes govern growth and division. Abnormally low activity of proteins that normally prevent cell division can result in cancer. Mutations in genes called **tumor suppressor genes** (discussed below) are responsible for such disease states. Abnormally high activity of proteins that normally stimulate cell division also can result in cancer. Chromosome rearrangements or mutations that produce oncogenes are responsible in these cases. Oncogenes, which can be carried by retroviruses, are discussed in the following section.

Cancer has been studied from several points of view. One has focused on understanding how chemicals called **carcinogens** introduce defects into cells. These compounds are often flat structures that can insert between the bases of DNA and cause errors during DNA replication. When cultured cells are treated with carcinogens, a small number keep multiplying rather than stopping when a monolayer is formed. These cells, which seem to ignore the signals from their neighbors to stop dividing, will cause tumors if injected into animals. Often the defects accumulate gradually, with the result that slowly growing tumors appear first, followed by fast-growing, potentially lethal ones.

Inherited forms of cancer have also been found. Examination of a type called retinoblastoma suggested that at least two defects are needed to elicit the cancerous state. Gradual progression, plus the requirement for multiple defects, led to the idea that some genes, now known as **tumor suppressor genes,** normally prevent pathological cell growth. One of the more extensively studied tumor suppressor genes is called *p53* (defects in *p53* have been associated with half of all human cancers). This gene appears to normally prevent cell proliferation when DNA has been damaged, presumably to allow time for repairs. Consequently, defects in *p53* result in the rapid accumulation of DNA damage that gradually leads to cancer. From this observation one would expect genes responsible for DNA repair to also be tumor suppressor genes. This is the case. In general, tumor suppressor defects act as recessive genes—one good copy is often enough to protect the cell, and most people are born with two copies. Over time, however, our cells gradually accumulate damage. When both copies of a tumor suppressor gene are inactivated, that cell moves one step closer to losing control over growth. Some persons are born with defects in tumor suppressor genes, and they are predisposed to developing cancer at a younger age than the general population.

ONCOGENES

As pointed out earlier, normal cells of most types stop dividing when they have completely covered the surface of a petri dish. Tumor viruses perturb the biochemical pathways that control the growth and division of cells, resulting in uncontrolled cellular growth. Consequently, when cultured cells are exposed to a tumor virus, the infected cells continue growing and dividing even when there is no more room on the surface of the petri dish. The infected cells grow on top of each other, forming clumps **(foci)** that can be easily observed using a low power microscope. In many respects foci are like bacterial colonies, and they are often used for the same type of manipulation. When transplanted into an animal, the infected cells will often develop into tumors.

Retroviruses are thought to cause tumors in two general ways. First, viral integration can occur near or within an important control gene in the host chromosome, raising the level of expression of that gene in a way that leads to uncontrolled cellular growth (Figure 12-7a). The probability of this happening is generally low because viral integration is not specifically targeted to these sites. Second, the virus itself may carry a gene whose product alters normal cellular control pathways (Figure 12-7b). These virus-borne "control genes" are called oncogenes; they tend to produce tumor cells at high frequency. Viral oncogenes are often defective copies of cellular regulatory genes, which in the normal form are called **proto-oncogenes.** The products of these genes generally encourage cells to replicate. Some are surface receptors that bind to growth factors, some are part of a pathway signaling the presence of growth factors, and some control cell division. Conversion of a proto-oncogene to an oncogene often creates an overactive protein or produces abnormally high levels of a normal protein. Thus mutations converting proto-oncogenes to oncogenes are dominant: it is necessary to alter only one of the two gene copies to make the gene tumorigenic. Often two different oncogenes must be present to generate the tumor state.

Oncogenes frequently interfere with the normal functioning of the viral genes responsible for producing new virus. Thus RNA tumor viruses are often defective and sometimes require "helper"

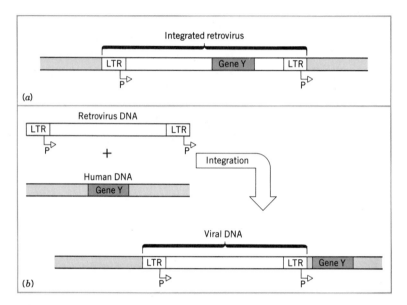

Figure 12-7 Stimulation of Gene Expression by Retroviruses. (a) Viral DNA contains a copy of an animal gene (gene Y), which comes under control of the strong viral promoter in the left LTR instead of a normal cellular promoter. Shaded regions indicate human DNA. **(b)** Insertion of a retrovirus immediately upstream from a human gene (gene Y) can place that gene under the control of the strong promoter in the right LTR of the virus. Shaded regions indicate human DNA.

viruses to reproduce. This may help explain why epidemics of cancer have not been observed. To date about 50 different oncogenes have been identified, and they have been found in about 20% of human tumors.

An important class of oncogene is called *ras*. A small family of related genes found normally in a variety of eukaryotic organisms ranging from yeasts to humans, the *ras* genes are associated with a variety of human and rodent tumors. Moreover, *ras* genes have been found as oncogenes in mouse and rat retroviruses. Nucleotide sequence comparisons between normal *ras* genes and oncogenic ones in retroviruses show that oncogenic *ras* genes contain mutations. It may be that some carcinogens cause cancer by chemically modifying the *ras* region of the DNA. Normal *ras* genes can also produce malignancy if the Ras proteins are produced in very large amounts. This can occur when a very strong promoter is placed immediately upstream from the gene. Retroviruses contain strong promoters in their LTRs, and the one in the 3' right-most LTR stimulates transcription outside the integrated virus (Figure 12-7*b*). Thus integration of the virus immediately upstream from a *ras* gene can lead to uncontrolled cell growth.

PERSPECTIVE

AIDS has brought retroviruses to almost everyone's attention. The disease was first clinically described in 1981. Six years later more than 50,000 Americans had been diagnosed with this condition, and many have since died. Another million were thought to be infected by the virus, and the number keeps climbing. It is now clear that transmission is accomplished via body fluids and that the disease is generally spread by sexual contact or exchange of blood and blood products. It has been difficult to develop a protective vaccine, and the disease cannot be cured by antibiotics, since these drugs generally work on bacteria, not viruses. Infection does, however, require several proteins unique to the virus, proteins such as reverse transcriptase, integrase, a specific protease, and the products of the viral regulatory genes. An attractive goal is to design chemical agents that attack these proteins but have minimal effects on the host cells and, hopefully, few undesirable side effects. The ability of HIV to evolve into a wide variety of different form, makes the development of vaccines and curative agents particularly challenging.

Formation of cancer cells is a complex process; indeed, we do not yet understand how normal cell division is controlled. However, the identification of oncogenes and tumor suppressor genes is progressing rapidly. Nucleotide sequence comparisons are revealing that certain mutant forms of *ras* and other gene families are associated with a predisposition to cancer. With genetic screening it is now possible to identify individuals whose genes contain some of the damaging mutations, and such persons can be counseled with respect to high-risk activities involving known carcinogens.

As the research effort intensifies to learn about AIDS and cancer viruses, our understanding of retroviruses in general will deepen considerably. This knowledge should in turn help us treat certain types of genetic disease. But the major impact of this research may be in the area of human engineering, for with an understanding of retroviruses comes the ability to use defective ones as cloning vehicles to insert genes into human chromosomes.

Questions for Discussion

1. How is AIDS spread?
2. How are the life cycles of bacteriophage lambda and the human immunodeficiency virus similar and how are they different?
3. Methods are being devised to block the replication of retroviruses to stop the spread of AIDS. For example, the drug AZT blocks reverse transcription, and other therapies are aimed at inactivating regulatory proteins such as Tat. Based on what you know about the life cycle of HIV-1, are these approaches likely to cure AIDS (i.e., to rid a person of the virus)?
4. If you were designing methods to find chemicals that would block stages in the AIDS virus life cycle, would you seek compounds that acted before or after integration?
5. What are some of the problems involved in using retroviruses as vehicles for inserting genes into humans?
6. Reverse transcriptase makes errors during the conversion of viral RNA to viral DNA, and persons infected with HIV-1 contain huge numbers of the virus. How do these two observations affect attempts to develop effective vaccines and therapies?

7. Persons infected with HIV-1 are particularly vulnerable to the airborne bacterial disease called tuberculosis, which is widespread in parts of the world where AIDS is spreading rapidly. Strains of the tuberculosis bacillus have been found that are resistant to most antibiotics. If the global spread of AIDS continues, what will be the public health implications of these two observations?

HUMAN GENETICS

Serial dilutions

1:10 100μl – 900μl
1:50 20μl – 980μl
1:250 4μl – 996μl
↙
How many times this
number fits in to
amount counted.

eg. 1000μl/250 = 4

∴ 4 : 996
dilution.

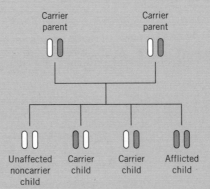

Carrier
parent

Carrier
parent

Unaffected
noncarrier
child

Carrier
child

Carrier
child

Afflicted
child

*Some genetic diseases are inherited
according to simple patterns.*

CHAPTER THIRTEEN

HEREDITY
Patterns of Inheritance and Genetic Instability

Overview _____

Human genetic information is stored in roughly 100,000 genes in DNA molecules that are organized into large structures called chromosomes. The chromosomes are structurally distinct, and microscopy has revealed that somatic (body) cells contain 22 pairs plus two sex chromosomes. One member of each chromosome pair originated from the individual's father and the other from his or her mother. Thus every person has two slightly different copies of the 23 distinct DNA molecules and therefore two copies of each gene in those molecules (males have only one copy of genes on the X and Y sex chromosomes). Genetic information passes from generation to generation in patterns that follow simple rules of probability, and for many hereditary traits the chance that a particular child will inherit a certain trait can be calculated. When the trait is caused by an alteration of a single gene, the analysis is comparable to that used for flipping coins. Much of our knowledge about human genes concerns genetic diseases, since the suffering associated with these traits has commanded attention from the medical community for many generations. By looking through family records, many of us can see evidence of genetic disease that warns us about our own future health.

INTRODUCTION

Heredity is the transmission of characteristics from parent to off-spring by means of genes. Earlier chapters described the chemical nature of genes, explaining how information is stored, accessed, and used. It is currently estimated that human DNA is subdivided into roughly 100,000 genes. Access to the information bound into many of these genes is controlled by the products of other genes. Elaborate networks of gene control exist, presumably to ensure that specific proteins are made in the correct amounts, at the proper time, and in the appropriate tissue. In spite of such complexity, simple patterns of inheritance can sometimes be seen. Being able to recognize those patterns in our own families gives us predictive power in a statistical sense. For example, we can know the odds that our children will possess a particular trait. While this type of information does not tell us whether a particular child will have the trait, it is becoming increasingly valuable for directing us toward specific DNA tests that will. This chapter focuses on patterns of inheritance that are commonly observed with disease-causing genes.

PATTERNS OF INHERITANCE

Many of the ideas behind the study of human genetics originated from work of Gregor Mendel, an Augustinian monk working in the 1860s. He discovered that inherited features are passed from one generation to the next as if they were particles. He asserted that the particles controlling heritable traits come in pairs and that progeny (offspring) receive one member of a pair from each parent. Mendel bred different varieties of peas in his monastery garden and then compared the characteristics of parents and offspring. When he mated tall plants with tall plants, he always got tall offspring. Like-wise, when he crossed short plants with short plants, he always obtained short plants. Thus the two types were "true breeding." But when he bred tall plants with short ones, he produced only tall offspring, not a mixture and not medium-sized plants. Tallness overpowered shortness; tallness was **dominant** to shortness. We can depict the situation symbolically by

TT (tall parent) $\times$ tt (short parent) $=$ Tt (tall offspring)

where T represents the dominant form of the particle (gene) and t represents a hidden, **recessive** form. The hidden nature of shortness became apparent when Mendel mated two of the Tt tall progeny plants. About a quarter of their offspring were short. Thus, the shortness trait must have been retained in the first mating. It was somehow covered up. After many crosses, Mendel discovered that the tall offspring of a Tt (tall) $\times$ Tt (tall) mating were not all the same: two out of three carried the concealed, recessive gene (t) for shortness. His numbers were most easily explained by the following scheme, in which offspring receive one trait (T or t) from each parent (indicated by subscripts 1 and 2):

Parents **Offspring**

T_1t_1 (tall) $\times$ T_2t_2 (tall) $=$ T_1T_2 (tall) $+$ T_1t_2 (tall) $+$ T_2t_1 (tall) $+$ t_1t_2 (short)

The distribution of characteristics among the progeny in this example follows the same rules used to describe the results from flipping two coins. (Let one coin represent the male parent, the other coin the female parent, heads the symbol T, and tails the symbol t. When both coins come up heads, one chance in four, we get the T_1T_2 situation. The t_1t_2 situation arises when both come up tails. T_1t_2 and T_2t_1 arise when one coin is heads and the other tails.)

Mendel further found that some traits are independent—the inherited particles (genes) behaved as if others did not exist. We now know that independent inheritance arises when genes are on different chromosomes, the long threadlike, protein–DNA structures that carry hereditary information. (Genes on the same chromosome will tend to stay together during the formation of germ cells and will not be inherited independently.) Since Mendel's time we've also learned that many complicated patterns of inheritance can occur. In some cases neither form of a gene is dominant, and matings give a "blended" result—one plant, called the four-o'clock, sometimes has white flowers and sometimes red; mating of white-flower and red-flower forms produces offspring with pink flowers. Other results arise from the effects of multiple genes. Nevertheless, extreme forms of some human diseases are inherited according to the simple rules derived by Mendel.

HUMAN CHROMOSOME INHERITANCE

Human somatic cells contain 46 chromosomes, 22 pairs plus two each of the sex chromosomes (XX in females and XY in males). Each member of a chromosome pair carries the same genes; thus, for most genes humans have two sets (exceptions occur in males, where the genes on the X and Y chromosomes are present in only one copy per cell). A process called **meiosis** reduces the chromosome numbers during the development of **germ cells** (**sperm** and **eggs**) so each germ cell contains only one member of each chromosome pair. Thus germ cells contain 23 chromosomes rather than 46. During fertilization, sperm and egg unite, producing a **zygote,** a cell that contains pairs of each chromosome type. Many cell divisions then follow, each involving faithful duplication of the chromosomes. In the resulting human being, every DNA-containing cell (except germ cells) has two copies of each gene except those on the X and Y chromosomes in males. Since one gene copy came from the father and one from the mother, the two copies usually contain differences in nucleotide sequence.

Although the germ cells formed in a given individual are all derived from cells having the same DNA information, each germ cell ends up with an information content that differs from that of all other germ cells. The differences are due in part to an exchange of portions of DNA between the paternal and maternal members of each chromosome type (Figure 13-1*a*). The exchange process, which is called recombination, involves a physical breaking and rejoining of DNA that occurs when each pair of chromosomes aligns during formation of germ cells. A second factor in creating germ cell diversity is chromosome sorting: a particular germ cell ends up with either the paternal or maternal copy of each of the 23 different chromosome types; Figure 13-1*b* shows the combinations possible with two chromosome pairs. This random sorting by itself can generate 2^{23} different germ cells. When added together, recombination and sorting make each germ cell unique. Thus each baby is unique. An exception occurs with identical twins—they are generated after

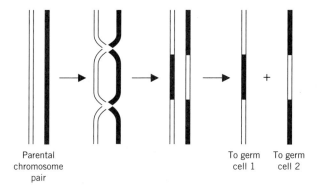

Parental
chromosome
pair

To germ
cell 1

To germ
cell 2

(*a*) Recombination

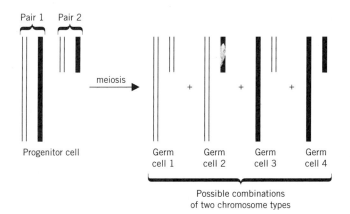

Pair 1   Pair 2

meiosis

Progenitor cell

Germ
cell 1

Germ
cell 2

Germ
cell 3

Germ
cell 4

Possible combinations
of two chromosome types

(*b*) Random sorting

Figure 13-1 Generation of Germ Cell Diversity. (a) Effects of recombination. During formation of germ cells, pairs of chromosomes align, and
portions of DNA are exchanged through a process of DNA strand breakage
and rejoining, which on average occurs 2 to 3 times per chromosome. Individual members of the pairs then end up in different germ cells. Thus the
nucleotide sequence of germ cell DNA is different from that found in the
cell that gave rise to the germ cell. **(b)** Random sorting of chromosomes.
Two pairs of chromosomes are shown in the somatic cell (progenitor cell)
that gives rise to germ cells. Meiosis results in only one member of each
pair of chromosomes in each germ cell, thus reducing the number of chromosomes by half. Chromosome sorting is random, so all possible combinations
arise in the population of germ cells.

255

fertilization has occurred, and so they have exactly the same DNA information.

Single-Gene Disorders: Autosomal Dominant

"Autosomal" refers to all chromosomes other than the sex chromosomes (X and Y), and "dominant" designates a trait that is manifested even when only one of the two genes specifies the trait (in Mendel's studies, only one copy of *T* was needed to generate a tall plant). Consequently, autosomal dominant diseases occur equally in males and females, and both sexes pass these conditions to their children. In classic cases, every affected individual will have at least one parent who is also affected (for pattern of inheritance, see Figure 13-2). Since these diseases are rare, an affected person usually has only one copy of the responsible gene in the disease-producing state and usually mates with someone having two normal gene copies. Thus, the chance for a child to be afflicted is usually 50% (Figure 13-2), high enough odds for the disease to show up in every generation. However, serious dominant disorders, which tend to strike adults,

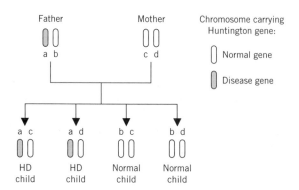

Figure 13-2 Inheritance Pattern of a Lethal Disease by Means of an Autosomal Dominant Gene. In this example, the chance that a child will inherit the gene for Huntington disease (HD), is 50%. (Diagram from K. Drlica, *Double-Edged Sword: The Promises and Risks of the Genetic Revolution*, © 1994 by Karl A. Drlica, with permission of Addison-Wesley Longman Publishing Co., Inc.)

can appear to skip a generation if an affected person dies of other causes before showing signs of the disorder. Huntington disease, a neurological disorder characterized by a loss of control of body movements, is an example of this type of malady.

From a biochemical perspective, one can view autosomal dominant disorders as arising from either insufficient activity of a normal gene product or creation of a harmful product. In the former situation, therapy may be possible by increasing the concentration of the product. Therapies using ribozymes (see Figure 11-6) may prove useful for eliminating a harmful product.

Single-Gene Disorders: Autosomal Recessive

Autosomal recessive diseases are clinically apparent only when the mutation is present in both copies of the responsible gene. Since by definition autosomal genes are not located on the sex chromosomes, males and females are affected in equal proportions. A parent who carries just one copy of the gene in the disease form generally appears normal. Such parents are called carriers because only their descendants can be affected. Since the probability that a child will be affected is only 25% (see Figure 13-3 for inheritance pattern), multiple occurrences in a family may not be seen in regions or cultures where small families are common. Consequently, this type of disease will appear sporadically. Cystic fibrosis, phenylketonuria, sickle-cell disease, and Tay–Sachs disease are examples of this type of disorder.

Many autosomal recessive diseases are due to biochemical defects in enzymes. It may be that because enzymes often provide almost normal function when present at half the normal level, such diseases are seen only when both copies of the gene are defective. Recessive disorders are generally diagnosed in children; there is hope that many can be cured by adding a normal copy of the gene to cells of afflicted organs.

Single-Gene Disorders: X-Linked

X-linked disorders arise from genes located on the X chromosome. Since males have only one X chromosome while females have two,

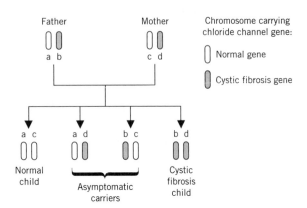

Figure 13-3 Inheritance Pattern of a Lethal Recessive Gene. Only one out of four or 25% of the children will inherit both copies of the gene in the diseased form. (Diagram from K. Drlica, *Double-Edged Sword: The Promises and Risks of the Genetic Revolution,* © 1994 by Karl A. Drlica, with permission of Addison-Wesley Longman Publishing Co., Inc.; adapted from M. A. McPherson and R. L. Dormer, *Molec. Aspects Med.* 12 : 4, 1991; Elsevier Science Ltd., Kidlington, UK.)

the clinical manifestations and inheritance patterns differ for males and females. An affected male can never pass the disorder to his sons because sons always receive the Y, not the X, from their fathers. However, an affected father passes his X chromosome to all his daughters, so they will all receive the mutant gene. The paragraphs that follow describe three common patterns of X-linked inheritance.

Recessive X-linked disorders are expressed primarily in males, since males have only one X chromosome and thus only one copy of that set of genes. Mothers and half of the sisters of affected boys will generally be carriers without symptoms, since they have two X chromosomes, one with a normal gene copy and one with an abnormal copy. When family trees (pedigrees) are examined, the disorder will be seen in maternal uncles of affected males (Figure 13-4) and in male cousins descended from sisters of the mother. Sons of carrier mothers have a 50% chance of having the disorder. This pattern is observed with hemophilia, a common X-linked dis-

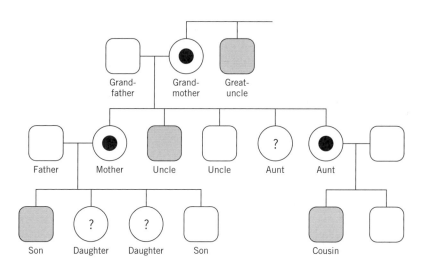

Figure 13-4 Inheritance Pattern of an X-Linked Recessive Disorder.
Squares indicate males, circles females. Solid symbols represent persons
showing disease symptoms. Asymptomatic carriers are indicated by a dot
in the circle; question marks refer to cases of undetermined carrier status.

ease characterized by the inability of blood to clot. Female offspring
are affected only when both X chromosomes carry the disorder. A
female hemophiliac must be the daughter of an affected man and a
woman who either is affected or is a carrier. Color blindness is one
of the X-linked recessive disorders common enough to show up
in women.

Dominant X-linked disorders appear in females as well as in males.
While an affected mother transmits the disorder to only half of her
daughters or sons, an affected father transmits it to all his daughters
but to none of his sons. Consequently, females tend to display this
type of disorder twice as often as males.

A third pattern arises when a dominant X-linked trait is lethal for
all males, since there can be no carrier father to pass along a mutant X
chromosome (all such males die before birth). Diagnostically, women
carrying this type of disorder have a high rate of spontaneous abortion of male fetuses.

X-linked disorders in females can be complicated by a phenomenon called **X-chromosome inactivation.** One of the two X chromosomes is normally inactivated in all **somatic** (body) cells of female embryos. This means that genes on only one of the two X chromosomes can be expressed and used as information for proteins. The complicating feature is that inactivation is random: in some cells the paternal X will be active while in other cells the maternal X will be active. Once a chromosome has been inactivated, it remains inactive in all cells arising from that particular embryonic cell. Thus, X-inactivation makes females genetic mosaics: their tissues contain clusters of cells (each derived from a single embryonic cell) in which one or the other X chromosome is active. The random nature of X-inactivation means that different females will have different clusters of cells affected by the disorders. Consequently, females can vary considerably in the manifestation of X-linked disorders.

Multifactorial Disorders

Many disorders "run in families" without a clearly predictable pattern. Among the more common diseases of this type are hypertension, gout, diabetes mellitus, and peptic ulcer disease. Some single-gene defects fail to cause noticeable disease unless other mutations or particular environmental conditions are also present. In a sense, a molecular threshold must be passed before disease is seen. This complexity makes multifactorial disorders difficult to study and predict.

New Mutations

New mutations can occur in any generation and can be inherited from then on. It is estimated that about one in 100,000 individuals acquires a mutation in a given gene. Many mutations have no harmful effect, however, so the frequency of clinically expressed, single-gene disorders from new mutations is much, much lower than 1 = 100,000. New mutations causing recessive disorders are difficult to identify (carriers don't manifest the disease); consequently, the new disease mutation may not be observed for generations. However, new dominant or X-linked disorders do arise often enough to be detected.

Chromosomal Disorders

As pointed out at the beginning of this book, chromosomes are diffuse, DNA-containing structures during most of the cell cycle. However, during the short period of the cycle when cells divide, chromosomes condense enough to be easily distinguished by light microscopy. At that stage a chromosome looks like a hot dog with a constriction. The location of this constriction is characteristic for each chromosome type. Prior to cell division, the chromosomes duplicate, and for a while the two daughter chromosomes remain attached. This gives the chromosomes an X-like appearance when examined by microscopy (Figure 13-5). Identification is possible because each chromosome type has a characteristic size and location of its constriction (centromere). In addition, chromosomes display characteristic staining patterns when treated with dyes.

Examination of many, many fetal cells has revealed that chromosomes are occasionally gained or lost, broken with loss of a piece, broken and rejoined to another chromosome, or deleted for one arm while duplicated for the other. Some of these events occur often enough to be correlated with particular diseases. For example, Down syndrome is associated with the presence of three rather than two copies of chromosome number 21. A variety of factors influence the frequency and types of chromosomal disorders, most of which are still undefined. It is clear, however, that these disorders increase with maternal age, making fetal chromosome examination highly useful for identifying these disorders in the pregnancies of older women. It is also likely that particular genes predispose some families to chromosomal disorders.

GENETIC INSTABILITY

Several diseases are characterized by increasing disease severity or progressively earlier age of disease onset as the disease-causing gene passes from one generation to the next. Such disease behavior was

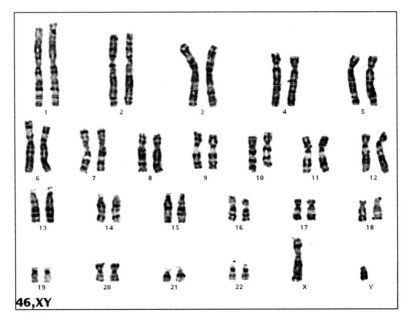

Figure 13-5 Metaphase Chromosomes. Pairs of human chromosomes, ordered and displayed to facilitate comparison. The banding patterns appear because some regions stain more strongly than others. (Photograph courtesy of Dr. Lillian Y. F. Hsu, Prenatal Diagnosis Laboratory of New York City.)

initially difficult to understand, since DNA is generally considered to be a stable carrier of genetic information: complementary base pairing to a template strand assures overall fidelity during replication, DNA polymerases often have editing activities to correct errors, and repair enzymes fix a variety of damages. Examination of nucleotide sequences in DNA from afflicted individuals revealed that particular genes contain short (trinucleotide) repeats whose number correlates with disease severity—expansion of the repeat number seemed to cause disease (Figure 13-6). In some extreme cases, repeat numbers become so unstable that they increase or decrease as body

Disease	Repeat Unit	Number of Repeats		Triplet Repeat Location
		Normal	Disease	
Myotonic dystrophy	CTG	5–37	50 to Thousands	3' side of one coding region, 5' side of another; effect unclear
Fragile X syndrome	CGG	6–42[a]	Thousands	5' side of coding region; may interfere with mRNA synthesis
Kennedy disease	CAG	17–26	40–52	In coding region of gene; probably produces aberrant protein
Huntington disease	CAG	11–34	37–86	In coding region of gene; probably produces aberrant protein

[a]50–200 repeats is considered to be a premutation.

Figure 13-6 Trinucleotide Expansion Diseases. The diseases listed are characterized by nucleotide triplets that are repeated many times in particular genes. With successive generations the number of repeats increases, as does the severity of the disease.

cells reproduce, giving an individual different numbers of trinucleotide repeats in different cells.

The trinucleotide expansion diseases tend to be neurological disorders that do not show up until adulthood. With fragile X syndrome, two events appear to occur during disease progression. First, a premutation arises that increases the number of trinucleotide repeats (according to population modeling estimates, the premutation for fragile X syndrome persists for about 60 generations). Then the number of repeats expands and becomes increasingly unstable—the longer the repeated region, the more it expands with every generation. Age of onset drops, and eventually the disease becomes so severe that it prevents reproduction. Then, of course, that line of the family ceases to pass on the disease.

PERSPECTIVE

By studying hereditary traits that are due to alterations in single genes, we have discovered that patterns of inheritance can be quite simple. Complications arise when traits are due to multiple genes. Over the years medical geneticists have catalogued more than 4000 human genetic disorders, and it is likely that each of us has nucleotide sequences we would like to change to improve our bodies. Recombinant DNA technology has made human DNA accessible to gene hunters, and we can expect more and more diseases to be associated with specific nucleotide sequences. Knowing that a particular disease runs in a family can be valuable to members of that family if corrective action can be taken. In such cases it will make sense for a family to participate in testing to identify individuals who could benefit, from knowing the information contained in their DNA. The next chapter focuses on DNA tests and other applications of molecular genetics to humans.

Questions for Discussion

1. How would your self-image have been different if you had been told as a child that you were a carrier of a serious genetic disease such as Tay–Sachs disease or sickle-cell disease? Do you think your parents would have treated you differently or allocated family resources to you differently if they had known you had this condition? (Note that carrier status is disease-free.)
2. Some families conceal or suppress evidence that a legal parent of a child may not be the biological parent. How could genetic tests uncover such a secret?
3. How do adoption, artificial insemination, and surrogate parenthood influence how you interpret your family tree with respect to inheritance of a genetic disease?

4. Using Mendel's experiment with peas as a model, prepare a chart to show how a mating between *Tt* (tall) parents gives rise to one *TT* (tall), two *Tt* (tall), and one *tt* (short) offspring.

5. What types of offspring would arise from a mating in which one parent is *Tt* (tall) and the other is *tt* (short)?

6. Hemophilia is a common X-linked disorder. Why can an affected male never pass the disease to his sons? Can one of his daughters suffer from the disease?

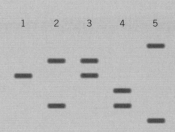

*People and their genes can be
tracked by DNA typing (fingerprinting).*

APPLICATION OF HUMAN GENETICS

The Genome Project, Genetic Testing, Gene Therapy, and DNA Fingerprinting

Overview

An extensive scientific effort is being undertaken to relate nucleotide sequences to human biology, an effort that will greatly expand the uses of genetic analyses. Already recombinant DNA methodology is allowing biologists to associate particular nucleotide sequences with specific diseases, and the combination of DNA analyses and family history makes it possible to predict some diseases years before symptoms appear. Thus prenatal DNA testing can give couples advance notice of disease conditions. In some cases selective abortion will be one way to avoid a family tragedy. If, however, DNA analysis reveals only a predisposition to disease, the information will allow a person to adjust his or her behavior to minimize the effects of the condition, or delay its onset. In a few cases efforts have been made to deliver normal genes to afflicted persons and change the course of a disease. Nucleotide sequence information also makes it possible to establish family relationships among people and to show with a high degree of certainty that a given person was present at the scene of a crime.

INTRODUCTION

As emphasized in Chapter 13, genes pass from one generation to the next according to simple rules that sometimes allow us to calculate the chance that a person will inherit a particular trait. We know, for example, that when both parents are asymptomatic carriers of the genetic disease called cystic fibrosis, a quarter of their children will be afflicted with the disease and half will be carriers. However, the calculations do not indicate in advance *which* child will become ill. Recombinant DNA technology makes it possible to identify individual humans as afflicted through the known association of specific nucleotide sequences with specific diseases.

The identification of disease-causing nucleotide sequences is also allowing the medical community to do something about genetic disease. For example, it is becoming increasingly common to identify and abort afflicted fetuses in families known to have a high frequency of cystic fibrosis, and preemptive surgery is becoming increasingly popular for persons genetically susceptible to certain types of cancer. Nucleotide sequence information will also make it possible to treat some genetic diseases. One approach utilizes recombinant DNA methods and bacteria to produce large, virus-free quantities of particular proteins that can then be injected into patients. This has been especially important for defects in hormones (human growth hormone, erythropoietin) and tissue factors (blood clotting factors), since these proteins are difficult to obtain from human tissues. Another approach involves removing cells from an afflicted person for insertion of normal gene copies, whereupon the engineered cells are returned to the person. The second strategy, if used to alter sperm and eggs, could lead to permanent genetic changes in a family. In a sense, humankind now has the power to change its own nature.

GENETIC AND PHYSICAL MAPS

The **human genome,** the genetic material (DNA) of human germ cells, comprises a nucleotide sequence about 3 billion base pairs long. One of the first steps in understanding how this genetic information is organized involves formulation of maps that relate specific features of the genome. As mentioned in Chapter 13, every person inherits two chromosome sets from his or her parents. The traits from genes on each chromosome tend to be inherited together because these genes are physically linked; that is, all the genes of one chromosome are in the same DNA molecule. Cells contain two copies of each chromosome, each containing the same genes in the same order (the nucleotide sequences usually are not identical, since one copy came from the mother and the other from the father). Such chromosomes are called homologous. Occasionally the two gene copies will switch chromosomes. As pointed out in Chapter 13 (Figure 13-1*a*), switching of nucleotides from one DNA molecule to another is the result of a natural DNA breaking–rejoining process we call recombination. Examination of recombination is the basis for construction of **genetic maps.** Recombination can be observed if two conditions are met: the two gene copies differ in some recognizable way (one copy might be defective and cause a disease), and inheritance is followed relative to another nearby gene that also has different forms on the two homologous chromosomes. Consider the example shown in Figure 14-1 in which two genes are represented by solid and open symbols, respectively. Assume that each gene has a dominant form (*A* for one gene, *B* for the other) and a recessive form (*a* for one, *b* for the other). Next suppose that you inherited the dominant *A* and *B* types from one parent and the recessive *a* and *b* types from the other parent (stage 2, Figure 14-1). Most of your children would inherit from you *A* and *B* together or *a* and *b* together (left pathway, stages 3 and 4, Figure 14-1). However, if recombination occurs between the solid and open genes (right pathway stage 3, Figure 14-1), some of your children would inherit *A* and *b* or *a* and *B* together. How often *Ab* and *aB* children arise depends on the nucleotide distance (*d*, Figure 14-1) between the solid and open genes. The farther apart they are, the greater the chance that recombination will occur between them. Geneticists use records of the frequency of *Ab* and *aB* children to estimate the distance between the genes that carry these characteristics (actual analyses have additional complexities because children inherit two copies of each gene, one from each parent).

By knowing the genetic distances among a number of genes, it is possible to construct a genetic map without having any nucleotide sequence information. An example for three genes is shown in Figure 14-2. Once the distance between genes *A* and *B* has been established,

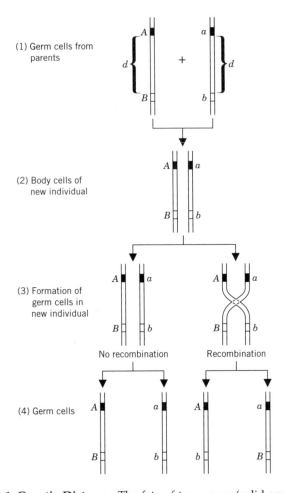

Figure 14-1 Genetic Distance. The fate of two genes (solid and open) on a pair of **homologous chromosomes** is shown as they pass from one generation to the next. **(1)** Germ cells from the parents combine to generate the new individual **(2)** without any rearrangements in the region being examined. During formation of germ cells in the new individual **(3)**, a process of DNA breakage and rejoining (recombination) can cause the traits labeled *B* and *b* to switch chromosomes. In this example four different types of germ cell **(4)** are produced by the new individual. *Ab* and *aB* germ cells are called recombinants. The frequency at which recombination occurs between two genes is greater for genes that are farther apart (greater values of *d*). Thus the percentage of offspring that are recombinant depends on the distance between the two genes with traits *A,a* and *B,b,* the chromosome types.

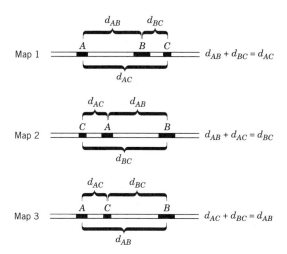

Figure 14-2 Genetic Mapping. Three arrangements of three genes, *A*, *B*, and *C*, are shown. Measurement of recombination frequency produces values for the relative distances *d* between these genes. The correct map order is established by determining which of the three equations shown at the right best fit the data.

the position of gene *C* can be determined by measuring *A-C* and *B-C* recombination frequencies. The same strategy is repeated with other genes located on the same chromosome.

The discovery of restriction endonucleases made it possible to map nucleotide sequence features to precise locations, since cutting of DNA takes place at precise sites (see Figure 7-5 for a simplified map in which restriction fragments are arranged). After blocks of nucleotide sequence have been determined for specific restriction fragments, the blocks can be ordered to generate large regions for which the sequence is known. Maps that locate nucleotide sequences are called **physical maps.**

During the late 1970s and early 1980s, polymorphic regions were discovered through variation in restriction fragment length. Each polymorphic region defined a specific spot in the genome. Restriction mapping was then extended on each side of the polymorphic regions to create local physical maps. In addition, the inheritance of individual polymorphic regions could be followed using electrophoretic and transfer hybridization analyses of DNA taken from parents and offspring. Thus polymorphic regions behave like genes, and genetic (recombination) analysis could be used to map the polymorphic regions near known genes (an example is shown in Figure 14-3). In this way physical and genetic maps are being aligned.

THE HUMAN GENOME PROJECT

By the mid-1980s it became clear that the availability of detailed maps, coupled with nucleotide sequence information, could greatly advance the study of genetic disease. For example, we could apply our extensive understanding of more primitive organisms to humans by finding genes in human DNA that correspond to known genes

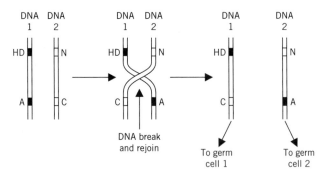

Figure 14-3 Genetic Recombination Between a Disease Gene and a Polymorphic Region. In the early 1980s analysis of large families revealed that the polymorphic region called G8 tends to be inherited with Huntington disease (HD). In the example, the HD gene was normally inherited with the A form of the polymorphic region. By following the inheritance of the polymorphic region, it was possible to predict which children were destined to suffer from this incurable neurologic disease. Occasionally (5% of the time), a child was born in whom the HD gene was associated with the C form of the polymorphic region. As a result of this gene switching, which was due to recombination between the polymorphic region and the HD gene, early DNA tests for Huntington disease were only 95% accurate. (Drawing adapted from K. Drlica, *Double-Edged Sword: The Promises and Risks of the Genetic Revolution,* © 1994 by Karl A. Drlica, with permission of Addison-Wesley Longman Publishing Co., Inc.)

of lower organisms. Even without extensive data banks, it has been possible to identify a human gene involved in cancer from knowledge of yeast genes. The National Research Council, recognizing the potential value of maps and nucleotide sequence information, recommended that a concerted effort be made to develop this type of information as a primary database for the study of human biology. This effort, funded by the National Institutes of Health and the U.S. Department of Energy, became known as the Human Genome Project. The economic importance was quickly recognized in other countries, and the project is now an international enterprise.

The initial phase of the genome project has focused on correlating physical and genetic maps. Large-scale physical mapping often involves cloning many pieces of human DNA and then arranging the cloned material into long contiguous regions called **contigs.** At the same time, many genes encoding specific proteins are being cloned using the amino acid sequence of the protein and the genetic code (Figure 2-6) to construct suitable oligonucleotide probes. Once cloned, the gene sequence is used to locate the gene on a particular chromosome and even within a particular contig (the individual clones provide hybridization probes that can be used to locate cloned regions).

GENETIC TESTING

"Genetic testing" is a general term that refers to a variety of processes that reveal whether a given individual has a particular trait. Three types of analysis are used for disease testing. One involves **cytological** examination of chromosomes obtained from fetuses to detect chromosome breaks, rearrangements, and incorrect numbers. A second involves analysis of blood for protein types known to be associated with particular genetic diseases. The third includes DNA tests for particular nucleotide sequences correlated with specific diseases.

DNA tests, which are usually carried out with families that have been repeatedly struck by a particular hereditary disease, involve extraction of DNA, either from blood cells or from cells grown in laboratory dishes, followed by examination of particular regions of DNA. When a specific nucleotide sequence alteration is thought to cause the disease, regions containing those sequences are amplified by the polymerase chain reaction, and the sequence is determined. In other cases, the nucleotide sequence responsible for the disease is unknown, but a particular form of a polymorphic (variable) region is known to be inherited at a high frequency with the disease gene. In these cases, DNA samples are obtained from other family members, some of whom are afflicted with the disease.

The family DNA samples can be examined in two ways. In one, the DNA is cut with particular restriction endonucleases, the resulting fragments are separated by gel electrophoresis, and the disease-associated RFLP is identified by transfer hybridization (see Figure 8-2). The fragment size reveals, to a high degree of certainty, whether the donor of the DNA sample is destined to suffer from the disease (see Figure 14-4). The second method involves analysis of CA-microsatellite DNA. Human DNA contains many regions in which simple sequences, such as CA, are repeated over and over. Such regions are called microsatellites, and for some of them the number of repeats varies from one person to the next, as illustrated later in Figure 14-5. When the nucleotide sequence outside the repeat is known, oligonucleotide primers can be constructed to use for PCR amplification of the microsatellite. When the amplified regions are reanalyzed by gel electrophoresis, unique sizes are detected that are inherited along with specific diseases, just as described for RFLP analysis.

Prenatal Testing

One application of genetic testing is in the identification of fetuses destined to be severely diseased. Prospective parents will then have highly reliable information on which to base their plans. One of the major tasks is to determine who should be tested and for what. For this, family histories are often valuable, as emphasized in Chapter 13. Recessive diseases tend to occur sporadically, however, and often a carrier family is not identified until after an afflicted baby has been born. In an effort to reduce this problem, adults are sometimes screened to find carriers and advise them of their status. A well-known example of adult screening concerns Tay–Sachs disease, a degenerative disorder that is especially common among Ashkenazic Jews. This effort, coupled with marriage restrictions and abortion, has reduced

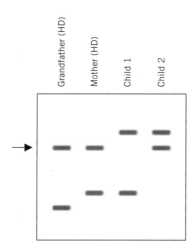

Figure 14-4 Predicting a Genetic Disease by RFLP Analysis of a Polymorphic Region. In this hypothetical example, DNA has been treated with a restriction endonuclease, the fragments have been separated by gel electrophoresis, and a polymorphic region (RFLP) has been identified by transfer hybidization using a probe specific to the polymorphic region. The DNA location of this particular RFLP is close to the gene responsible for the disease (during inheritance of the disease, this RFLP and the disease are inherited together at least 95% of the time because recombination occurs very rarely between them). The grandfather and mother are known to suffer from the dominant disease (labeled HD), and their DNA contains the RFLP form indicated by the arrow. The DNA from child 1 does not contain that band, and so that child is unlikely to suffer from the disease. DNA from child 2, however, does have the band indicated by the arrow, and that child has greater than 95% chance of being afflicted later in life.

the Tay–Sachs incidence by over 90%. Moreover, technological advances have now made it possible for Tay–Sachs couples to avoid having afflicted children without resorting to abortion. Egg cells are removed from a woman and fertilized with sperm in laboratory dishes. When the resulting embryos reach the eight-cell stage, one cell is separated from each embryo for DNA analysis. That indicates which embryos are free of the disease. Disease-free embryos are then implanted into the woman's uterus for further development.

Newborn Screening

Large numbers of newborns are routinely screened for a variety of genetic diseases, several of which respond to dietary treatment. One of the best examples is phenylketonuria (PKU), a disease caused by an inability of the body to break down the amino acid phenylalanine, high levels of which are often associated with mental retardation. Many of the disease symptoms of PKU can be alleviated by a diet that restricts the intake of phenylalanine. Another example is sickle-cell disease. Infants with this disease are prone to infection; if doctors are alerted, preventive antibiotic therapy can be very helpful.

It is important to emphasize that newborn screening is appropriate only with respect to diseases for which therapies are available. It is unwise to test newborns for incurable diseases or to help parents make reproductive choices for subsequent fetuses, since unfavorable results of genetic testing of a newborn can place a stigma on the child. Moreover, no child can give informed consent for genetic tests, and there are serious ethical problems associated with exposing nonconsenting subjects to the possibility of being linked to incurable diseases.

Testing for Adult-Onset Diseases

Many new tests are becoming available to help adults anticipate, and hopefully delay, common diseases usually associated with aging. Such ailments, which include artery disease and cancer, tend to be influenced by complex networks of genes that interact in subtle ways. Sometimes a genetic defect makes a person particularly susceptible to dietary or environmental factors that then lead to disease. Upon becoming aware of the susceptibility, the affected person can take preventive action.

One of the better understood examples is coronary artery disease. Blood **cholesterol** levels are often high in heart attack victims, and the arteries that supply blood to the heart muscle become clogged with cholesterol-containing deposits. One of the problems lies with a type of protein package called LDL (low density lipoprotein), which transports cholesterol from one organ to another. When combined with cholesterol, LDL has been called "bad cholesterol" because of the correlation of high levels of this substance with heart attacks. Members of some families tend to have high levels of LDL–cholesterol, and in such families there is a high incidence of heart attack. This observation led to a genetic study that identified a gene involved in coronary heart disease. The gene for familial hypercholesterolemia (FH) encodes a receptor protein that normally sits on the surface of some cholesterol-producing cells. There it acts as a sensor for LDL–cholesterol levels. When blood cholesterol levels are high, many of the receptors bind to LDL–cholesterol and signal the cells to stop making cholesterol. If the FH gene is defective, the high cholesterol signal is not received effectively, and the body produces too much cholesterol. When one copy of the gene malfunctions, blood cholesterol concentration is frequently two to three times above normal. Individuals with this condition often experience heart attacks by age 50. A defect in both copies of the gene leads to a six- to eightfold increase in blood cholesterol concentration. In these cases, heart attacks often occur during the teenage years.

Persons known to carry defective genes involved in cholesterol regulation can delay artery problems through medication and diet. Gene therapy methods, which are now under development, should further reduce LDL–cholesterol levels by delivering engineered liver cells to patients. Thus we are beginning to gain control over some forms of artery disease.

Cancer is another type of disease in which genetic predisposition is widespread. By the time of adulthood, most human cells slow their growth and multiplication, presumably because a set of proteins is made that blocks cell division. At the same time, mutations occur in the DNA molecules of the cells, and occasionally a mutation occurs in a tumor suppressor gene. Over time, defects accumulate in five or six of these genes, which function to encode proteins involved in blocking cell division, and progressively release the cells from normal growth control. This deregulation leads first to a benign growth and later to a highly invasive cancer. From this perspective cancer is a multihit, somatic (body) cell disease.

Sometimes a defective tumor suppressor gene is inherited; the defect is present in the germ cell before conception of the individual. Inevitably, then, all cells in the body of the individual will contain the defect. As a consequence of this lack of protection against uncontrolled cell growth, an abnormally small number of additional mutations are needed to result in cancer, and the individual is said to have a predisposition for cancer. Since tumor suppressor genes can be tissue-specific, inherited predisposition can apply only to certain tissues.

As pointed out in Chapter 12, one of the key anticancer genes is called *p53*. The protein product of this gene appears to participate in blocking DNA replication when DNA becomes damaged. Thus defects in *p53* allow damaged cells to proliferate, but defective *p53* does not by itself cause cancer. So far, defective *p53* has been associated with more than 50 forms of cancer, including a high percentage of colon, lung, breast, cervical, urinary tract, and bladder cancers. A form of skin cancer has even been associated with specific nucleotide changes in the *p53* gene. Once specific defects of particular genes have been associated with specific cancers, the degree to which a particular tissue is cancer prone can be assessed by means of nucleotide sequence analysis of DNA. Such tests will greatly increase the accuracy of diagnosis of any suspicious growth. For example, tests can be devised to determine whether cells sloughed off into the urine have defects in *p53* and in other genes associated with bladder cancer. That will make early detection of bladder cancer straightforward. The same principle should apply to any organ that releases easily collected cells, such as the esophagus and the uterus.

DNA tests can also indicate whether a particular person is cancer prone. For example, women in certain families are highly susceptible to breast cancer (their chance of being diagnosed by age 50 is 60%, versus 1% for the general population). Analysis of families that have experienced a high incidence of breast cancer has associated the cancer predisposition with the presence of an altered gene called *BCRA1*. These high-risk families can now use DNA tests to determine which of their members are particularly prone to breast cancer.

GENE THERAPY

Gene therapy is a process by which DNA is added to cells to overcome the disease consequences of a defective gene. The first authorized gene therapy experiment involved adding a gene for adenosine deaminase (ADA) to blood cells of a child who lacked the normal gene for ADA (ADA-deficient children suffer from a severe immune deficiency and have difficulty fighting normally innocuous microorganisms). Blood cells (lymphocytes) were removed from the patient and infected with a retrovirus (Chapter 12) into which a normal version of the defective gene had been inserted. Then the infected cells were injected into the patient. This treatment appeared to reduce the immune deficiency symptoms. Several other conditions, including cystic fibrosis, cholesterol imbalances, and certain types of cancer, are being attacked with comparable DNA technologies.

Before gene therapy can be widely used for many different diseases, several technical problems must be solved. First, it is necessary to identify and clone the gene responsible for the disease in question, since only then can a good copy be inserted into the patient. Of the thousands of genetic diseases, only a few percent have been attributed to specific genes. It is expected that results from the Human Genome Project will greatly speed gene identification and thus gene therapy.

Second, more effective methods need to be developed to deliver genes. Defective blood cells and skin cells can be removed from the body, infected with an engineered virus, and then returned in working form to the body. However, cells from most other tissues and organs, such as brain, do not function properly when returned to the body. For this reason, we can expect better methods to be developed for handling cells outside the patient's body.

We can also expect specialized viruses to be designed for carrying new genes directly to target cells inside the body. An example of virus technology is emerging from work on cystic fibrosis. A virus called adenovirus, one of the causes of the common cold, infects only respiratory tissue. Therefore, efforts are under way to use this virus to deliver a normal copy of the cystic fibrosis gene to lungs, a good first approach since lung damage is often the cause of death in cystic fibrosis.

A third technical problem with gene therapy concerns the short life span of certain cell types. T lymphocytes, for example, live only a few months. In the ADA gene therapy case, the gene treatment resupplies new, engineered T lymphocytes as the old ones die. A major advance would be the ability to engineer and deliver **stem cells,** the long-lived cells that divide to form T lymphocytes. These cells are hard to obtain from older children, but they can be recovered from umbilical cords. Thus there is hope that newborns with ADA deficiency can be cured by gene therapy using cells from their own umbilical cords.

A major ethical issue associated with gene therapy concerns genetic changes that are transmitted from patient to offspring, an event that can occur when genetic manipulation affects germ cells, that is, sperm or eggs. A special name, **germ line gene therapy,** has been applied to this type of procedure to emphasize its importance—it has the potential to gradually change the human race in a *directed* way. Among the philosophical questions arising are whether children have the right to inherit unmanipulated genomes and whether conventions of informed consent should apply to individuals who are not yet conceived.

Germ line gene manipulation, which is practiced extensively with mice, is already a refined technology. Fragments of DNA are injected directly into fertilized mouse eggs, whereupon some of the fragments become incorporated into chromosomes. The baby mice contain a new gene in every cell, including sperm cells and egg cells. The engineered mice then mate, and the new genes pass to the next generation of baby mice. Several genetic diseases of mice, including growth and fertility problems, have been cured this way. Although the methods are still imperfect, the basic technology could be applied to humans.

TRACKING PEOPLE

Identifying people has long been an important activity in our society. We want to know who the real father is, whether an immigration claim of kinship is true, or whether the accused is really the rapist. For many years identification efforts have rested on blood type similarities. However, the blood techniques were never very good because there are only a few different types of blood and because blood samples tend to decompose quickly. Our understanding of DNA now gives us a strong identification method. Unlike blood groups, there are as many distinguishable DNA types as there are people (except identical twins). Moreover, sufficient DNA can be obtained from a wide variety of body parts in addition to blood samples (a single hair root and nail clippings, to name just two).

The **DNA fingerprinting** method (also called DNA typing) is based on the observation that human DNA contains polymorphic regions in which very short, adjacent nucleotide sequences repeat over and over (up to 30 times). Such regions are called **VNTRs**, for Variable Number of Tandem Repeats. The number of repeats for a particular region of repeats, that is, a particular VNTR, varies from person to person. Consequently, if DNA were cut at specific points on each side of the VNTR, the distance between the cuts in the DNA would also vary from person to person; likewise, the length of the DNA fragment generated by such cuts would vary (Figure 14-5). Since a VNTR is a particular type of restriction fragment length polymorphism, the DNA/movie film analogy used in Chapter 7 can be used again to explain the idea. A VNTR is like a commercial inserted into a television movie, but in this particular type of commercial the same brief message is repeated many times during the advertising break. Local stations will use a given advertiser's brief,

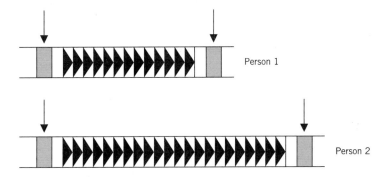

Figure 14-5 Variable Numbers of Repeated Segments Create Different Restriction Fragment Lengths. There are regions in human DNA that vary in the number of short, repeated regions. Samples of DNA from persons 1 and 2 contain 13 and 22 repeats, respectively. Outside these repeats are sequences (shaded) that can be used to construct oligonucleotide primers for amplification of the repeat region by PCR. Alternatively, the shaded regions can be recognized as cutting sites by particular restriction endonucleases. Amplification or cutting at specific sites (arrows) produces fragments of different lengths for the two people. (Drawing adapted from K. Drlica, *Double-Edged Sword: The Promises and Risks of the Genetic Revolution,* © 1994 by Karl A. Drlica, with permission of Addison-Wesley Longman Publishing Co., Inc.)

message repeatedly, but the number of repeats will differ from station to station. Thus the television version of the movie will vary slightly from one channel to another, just as a VNTR varies from person to person.

In practice, technicians first remove DNA from tissue samples and cut it with a restriction endonuclease to produce fragments of discrete length. For this, an endonuclease is chosen that cuts outside the VNTR. The lengths of the DNA fragments from different samples are then compared by gel electrophoresis and transfer hybridization. Since treatment with an endonuclease produces cuts at many sites in addition to those flanking a particular VNTR, the many cuts produce hundreds to thousands of *different* fragments of different sizes; only one or two would be from the particular VNTR of interest (in the analogy with film, the cutting would occur throughout the film as well as on each side of the commercial of interest). Transfer hybridization (Figure 8-2) is used to distinguish the VNTR band from the many other bands (the data would look similar to the sketch shown in Figure 14-4).

Since human chromosomes come in pairs, there are two copies of every VNTR. Occasionally the copies are identical, and then only one band will show up after transfer hybridization. Usually the two copies differ in repeat number, so the probe attaches to two bands. Forensics laboratories have spent several years testing collections of DNA probes used to detect particular VNTRs. Blood samples have been taken from hundreds of people, and company scientists have estimated for each probe (and thus each VNTR) how often the different DNA fragment lengths occur among people. From these estimates, odds have been obtained for how frequently a sample of DNA from any given person would match DNA from another person who was not closely related to the first. Consider, for example, a particular VNTR in which only 10% of the population has DNA with 22 repeats (person 2 in Figure 14-5). Then the chance that a random person in the population would also have DNA with 22 repeats is 1 in 10 (10%); the chance that a DNA sample from a crime scene would coincidentally have 22 repeats in this region is also only 1 in 10. Thus, if DNA from a crime scene and DNA from person 2 both had 22 repeats in the VNTR being examined, 90% of the population, but not person 2, would be eliminated from consideration.

When several different variable regions (VNTRs) are used in the analysis, the astounding power of DNA fingerprinting becomes obvious. For example, if two DNA samples have fragments that match for two VNTRs (a two-region match) and if the odds for the first VNTR to match were 1 in 10 and the second 1 in 20, then the chance that both VNTRs would match coincidentally is only 1 in 200 (the calculation is made by multiplying the two odds). Likewise, when five regions of DNA (five VNTRs) match, the chance that this has happened by coincidence is the product of the five individual matches. That number can be so small that almost everyone else is excluded from consideration.

PERSPECTIVE

Access to the complete set of human gene information will make molecular genetics the basis for the prevention and treatment of many, many ailments. But with these changes in medicine will come new social dilemmas. The immediate problems of prenatal (fetal) genetic screening, especially when coupled with abortion, revolve around ethnic diversity: genetic diseases are distributed unevenly among groups of people. For example, cystic fibrosis is prevalent among the Scots and English, PKU among the Irish and Polish, Tay–Sachs among Ashkenazic Jews, and sickle-cell disease among people of African and Mediterranean descent. Consequently, different screening programs are appropriate for different groups of people. Determining whether a particular screening program is worth the effort and expense will be a value judgment made by the controlling groups. When the controlling group is different from the target group, genetic screening programs can lead to abuse, mistrust, and accusations of genocide. In contrast, having the program administered by the target group greatly increases its chance of success, as illustrated by the successful Tay–Sachs program developed in Jewish communities for Jewish communities. Thus ethnic relationships within societies will strongly influence screening programs, and vice versa.

As screening programs become more sophisticated, the definition of "birth defect" will broaden: some couples will want to use screening/abortion to strengthen their families both physically and mentally. At the same time, other advances in molecular biology will improve the length and quality of life for those born with genetic disease. For example, new antibiotics have reduced the severity of the lung infections characteristic of cystic fibrosis, and many afflicted individuals now live nearly normal lives into their thirties. Babies born today with cystic fibrosis may live into their fifties. Comparable statements can be made about hemophiliacs, PKU babies, and sickle-cell infants. The combination of easy prenatal detection plus partially effective, but expensive therapies will present the medical ethics community, the insurance industry, and governments with a new problem: Should screening/abortion be encouraged to hold down health care costs?

The situation will become even more complex as segments of society begin to require genetic disclosure. Life insurance companies already consider certain genetic factors when issuing policies, and they will certainly want to take advantage of new information. Perhaps insurers will ask a person to produce genetic data to prove that he or she is a good risk. Genetic disclosure may also be required for certain occupations. For example, the public may demand evidence that candidates for high office have no genetic predisposition for mental illness; potential airline pilots or bus drivers may have to show that they are not predisposed to certain acute, hidden heart ailments.

The prospect of handicapped offspring may eventually block potential matings, either from choice or from societal pressures. For example, members of a group of Orthodox Jews decided to eliminate recessive disorders such as cystic fibrosis, Tay–Sachs, Canavan, and Gaucher diseases from their population. Their strategy consists of assigning community leaders to visit high schools having high concentrations of Orthodox students and inform them about the DNA screening program. Couples planning to date are encouraged to undergo tests and return for interpretation of the results, which will tell them whether they are genetically incompatible. This experiment should be watched closely, since it could be a window on the future. In particular, we need to see how well confidentiality is maintained, how a society deals with diseases that sometimes have only minor symptoms, and what happens to "genetic wallflowers"—persons who cannot find spouses because of genetic tagging.

In conclusion, the genetic revolution is beginning to provide us with many new ways to get the most out of our bodies. How each of us takes advantage of the new genetics depends on the precise information stored in our DNA molecules. While it is still too early to read your own program, some very valuable insights can be obtained by examining family medical histories. What you learn should be kept confidential, however: bear in mind that your relatives may not want to know what is likely to be in store for them, and accessibility to medical histories has occasionally led to discrimination.

Questions for Discussion

1. How can PCR be used in forensic DNA fingerprinting?
2. Could a small, homogeneous nation use recombinant DNA technology to improve the mental and physical powers of its population?
3. What factors do opposing attorneys argue about when DNA fingerprinting data are introduced in criminal trials?
4. Should society regulate mating between persons known to be carriers of genetic diseases for which therapies are lacking?

ADDITIONAL READING

Additional readings are listed for students who desire more information. Many of the articles are taken from *Scientific American* because of the high quality and wide availability of this publication. To facilitate finding relevant works, the articles are first listed topically. Complete entries follow in alphabetical order.

Topical Listing of Additional Reading

AIDS
    Gallo, 1986
    Gallo, 1987
    Greene, 1993
    Hazeltine and Wong-Staal, 1988
    Laurence, 1985
    Matthews and Bolognesi, 1988
    Mills and Masur, 1990

Behavior
    Greenspan, 1995
    Kandel and Hawkins, 1992
    Scheller and Axel, 1984

Cancer
    Bishop, 1982
    Cavenee and White, 1995
    Croce and Klein, 1985
    Hunter, 1984
    Sachs, 1986
    Weinberg, 1983

Cells
    Bretscher, 1987 (movement)
    Murray and Kirschner, 1991 (cell cycle)
    Welch, 1993 (response to stress)

Development
    DeRobertis, Oliver, and Wright, 1990

DNA replication and recombination
    Radman and Wagner, 1988
    Stahl, 1987
    Varmus, 1987
    Wang, 1982

Forensics
    Neufeld and Coleman, 1990

Gene expression
    Cech, 1986
    Chambon, 1981
    Cohen and Hogan, 1994
    Darnell, 1983
    Grunstein, 1992
    Holliday, 1989
    Nomura, 1984
    Ptashne, 1989
    Ptashne, Johnson, and Pabo, 1982
    Ross, 1989
    Sapienza, 1990
    Steitz, 1988
    Tjian, 1995
    Wang, 1982
    Weintraub, 1990

Genetic diseases
    Atkinson and Maclaren, 1990 (diabetes)
    Drlica, 1994 (general aspects)
    Geever et al., 1981 (sickle-cell disease)
    Lawn and Vehar, 1986 (hemophilia)
    Nathans, 1989 (color blindness)
    Rennie, 1994 (testing)
    Verma, 1990 (gene therapy)
    White and Lalouel, 1988 (chromosome mapping)

Genetic engineering/recombinant DNA
    Anderson and Diacumakos, 1981
    Capecchi, 1994
    Chilton, 1983
    Freifelder, 1978
    Gilbert and Villa-Komaroff, 1980
    Grobstein, 1979

Maniatis et al., 1978
Mullis, 1990
Murray and Szostak, 1987
Verma, 1990
Watson and Tooze, 1981
Weinberg, 1985

Immune system
Ada and Nossal, 1987
Boon, 1993
Capra and Edmundson, 1977
Collier and Kaplan, 1984
Ding, Young, and Cohn, 1988
Edelman, 1989
Godson, 1985
Grey, Sette, and Buus, 1989
Janeway, 1993
Kennedy, Melnick, and Dreesman, 1986
Leder, 1982
Lichtenstein, 1993
Marrack and Kappler, 1986, 1993
Mitchison, 1993
Nossal, 1993
Paul, 1993
Rennie, 1990
Smith, 1990
Steinman, 1993
Tonegawa, 1985
Weissman and Cooper, 1993
Wigzell, 1993

Industrial applications of microbes
Brierley, 1982

Infectious diseases (also see AIDS and Viruses)
Donelson and Turner, 1985 (trypanosomes)
Godson, 1985 (trypanosomes)
Prusiner 1995 (prions)

Intracellular structures
Grivell, 1983 (mitochondrial DNA)
Hakomori, 1986 (membranes)
Kornberg and Klug, 1981 (nucleosomes)
Lake, 1981 (ribosomes)
Unwin, 1984 (membranes)

Nucleic acids
    Cech, 1986
    Cohen and Hogan, 1994
    Darnell, 1983, 1985
    Dickerson, 1983
    Federoff, 1984
    Felsenfeld, 1985
    Grivell, 1983
    Paabo, 1993
    Rennie, 1993
    Watson, 1980 (*The Double Helix*)
    Weintraub, 1990

Proteins
    Doolittle, 1985
    Doolittle and Bork, 1993
    McKnight, 1991
    Prusiner, 1995
    Richards, 1991
    Rhodes and Klug, 1993
    Unwin, 1984

Viruses and antiviral agents
    Hirsch and Kaplan, 1987
    Johnson et al., 1994
    Jonathon, Butler, and Klug, 1978
    Pestka, 1983
    Ptashne, Johnson, and Pabo, 1982
    Simons, Ganoff, and Helenius, 1982
    Tiollais and Buendia, 1991
    Varmus, 1987
    Winkler and Bogel, 1992

Alphabetical Listing of Additional Reading

Ada, G. L., and Nossal, S. G. (1987) The Clonal-Selection Theory. *Scientific American*, August, pp. 62–69.
    A historical perspective of the development of our understanding of the immune system.

Anderson, W. F., and Diacumakos, E. G. (1981) Genetic Engineering in Mammalian Cells. *Scientific American*, July, pp. 106–121.
    Strategies for using bacteria and recombinant DNA techniques to engineer mammalian cells.

Atkinson, M. A., and Maclaren, N. K. (1990) What Causes Diabetes? *Scientific American*, July, pp. 62–71.
The relationship of the immune system to diabetes.

Bishop, J. M. (1982) Oncogenes. *Scientific American*, March, pp. 80–92.
Oncogenes cause cancer. They were first reported in viruses, but they have also been found in normal cells. Abnormal expression of these genes can lead to cancerous growth.

Boon, T. (1993) Teaching the immune system to fight cancer. *Scientific American*, March, pp. 82–89.
Tumor cells have unique antigens that can be used to elicit an immune response. This is a medical application of basic science.

Bretscher, M. S. (1987) How Animal Cells Move. *Scientific American*, December, pp. 72–90.
Animal cells other than sperm move by bringing into the cytoplasm pieces of the cell membrane that are later recycled to the surface.

Brierley, C. L. (1982) Microbiological Mining. *Scientific American*, August, pp. 44–53.
A discussion of the use of a bacterium called *Thiobacillus* in leaching copper from low-grade ore.

Capecchi, M. R. (1994) Targeted Gene Replacement. *Scientific American*, March, pp. 52–59.
The creation of mice bearing mutations in any known gene.

Capra, J. D., and Edmundson, A. B. (1977) The Antibody Combining Site. *Scientific American*, January, pp. 50–59.
Antibody structure is discussed, and the biochemical details of antigen–antibody binding are described.

Cavanee, W. K., and White, R. L. (1995) The Genetic Basis of Cancer. *Scientific American*, March, pp. 72–79.
A description of cancers associated with an accumulation of genetic defects.

Cech, T. R. (1986) RNA as an Enzyme. *Scientific American*, November, pp. 64–75.
Self-processing of RNA is described, a concept that has led to the development of ribozymes.

Chambon, P. (1981) Split Genes. *Scientific American*, May, pp. 60–71.
Some of the experimental evidence leading to the concept of introns and exons.

Chilton, M.-D. (1983) A Vector for Introducing New Genes into Plants. *Scientific American*, June, pp. 51–59.
Some plant tumors are caused by a bacterium that can be used to introduce genes into plants.

Cohen, J. S. and Hogan, M. E. (1994) The new genetic medicines. *Scientific American*, December, pp. 76–82.

Antisense and triple-helix strategies for controlling gene expression.

Collier, R. J., and Kaplan, D. A. (1984) Immunotoxins. *Scientific American,* July, pp. 56–71.
    Chemical modification of antibodies may lead to the development of a class of antitumor agent.

Croce, C. M., and Klein, G. (1985) Chromosome Translocations and Human Cancer. *Scientific American,* March, pp. 54–60.
    The movement of oncogenes and their activation can be traced by chromosome mapping methods.

Darnell, J. E. (1983) The Processing of RNA. *Scientific American,* October, pp. 90–101.
    In higher cells RNA is modified in a number of ways before it reaches the ribosomes in the cell cytoplasm.

Darnell, J. E. RNA. (1985) *Scientific American,* October, pp. 68–87.
    The various roles played by RNA.

DeRobertis, E. M. Oliver, G., and Wright, C. V. E. (1990) Homeobox Genes and the Vertebrate Body Plan. *Scientific American,* July, pp. 46–52.
    The homeobox genes are similar in many animals. Their involvement in the control of development of limbs and organs is discussed.

Dickerson, R. E. (1983) The DNA Helix and How It Is Read. *Scientific American,* December, pp. 94–111.
    A detailed treatment of DNA structures.

Ding, E., Young, J., and Cohn, Z. A. (1988) How Killer Cells Kill. *Scientific American,* January, pp. 38–45.
    Killer lymphocytes are the commandos of the immune system. They secrete protein molecules that punch holes in cells, causing them to leak to death.

Donelson, J. E., and Turner, M. J. (1985) How the Trypanosome Changes Its Coat. *Scientific American,* February, pp. 44–51.
    The causative agent of sleeping sickness evades the host immune response by switching on new genes that encode the surface protein of the parasite.

Doolittle, R. F. (1985) Proteins. *Scientific American,* October, pp. 88–99.
    Protein structure and enzyme action.

Doolittle, R. F., and Bork, P. (1993) Evolutionarily Mobile Modules in Proteins. *Scientific American,* October, pp. 50–56.
    An emerging pattern that may help explain the spread of modular elements of proteins during evolution.

Drlica, K. (1994) *Double-Edged Sword.* Addison-Wesley, Reading, MA, 242 pp. (paperback, 1996)
    The opportunities and risks of the genetic revolution are described using case histories; practical issues are considered. Volunteer agencies that focus on particular genetic diseases are listed.

Edelman, G. (1989) Topobiology. *Scientific American,* May, pp. 76–88.
Topobiology, the study of place-dependent interactions, may be revealing information about the origin of the immune system.

Fedoroff, N. V. (1984) Transposable Genetic Elements in Maize. *Scientific American,* June, pp. 84–98.
An introduction to mobile genetic elements in plants.

Felsenfeld, G. DNA. (1985) *Scientific American,* October, pp. 58–67.
DNA is flexible, and along its length its structure can change in ways that are probably important for gene control.

Freifelder, D., Editor (1978) *Recombinant DNA.* W. H. Freeman & Company, San Francisco, 147 pp.
A collection of articles from *Scientific American* that provides background material on the development of recombinant DNA technology.

Gallo, R. C. (1986) The First Human Retrovirus. *Scientific American,* December, pp. 88–101.
This, the following paper by the same author, and the paper by Hazeltine and Wong-Staal lay a foundation for understanding the AIDS virus.

Gallo, R. C. (1987) The AIDS Virus. *Scientific American,* January, pp. 46–73.
This lays part of the foundation for understanding the AIDS virus.

Geever, R. F., Wilson, L. B., Nallaseth, F. S., Milner, P. F., Bittner, M., and Wilson, J. T. (1981) Direct Identification of Sickle-Cell Anemia by Blot Hybridization. *Proceedings of the National Academy of Sciences U.S.A.* 78:5081–5085.
How sickle-cell anemia can be diagnosed using DNA methods.

Gilbert, W., and Villa-Komaroff, L. (1980) Useful Proteins from Recombinant Bacteria. *Scientific American,* April, pp. 74–94.
Recombinant DNA techniques are described with special reference to the procedures used to clone an insulin gene.

Godson, G. N. (1985) Molecular Approaches to Malaria Vaccines. *Scientific American,* May, pp. 52–59.
The proteins of the outer coat of the malaria parasite appear to serve as decoys to deflect the host immune system.

Greene, W. (1993) AIDS and the Immune System. *Scientific American,* September, pp. 99–105.
A description of the human immunodeficiency virus, how it replicates, and how it interacts with the immune system.

Greenspan, R. J. (1995) Understanding the Genetic Construction of Behavior. *Scientific American,* April, pp. 72–78.
A description of fruit fly experiments in which changes in genes affect fly behavior.

Grey, H., Sette, A., and Buus, S. (1989) How T Cells See Antigen. *Scientific American,* November, pp. 56–64.

T cells are an important part of the immune system; their mode of interaction with other cells of the immune system is described.

Grivell, L. A. (1983) Mitochondrial DNA. *Scientific American,* March, pp. 78–89.
Mitochondria are subcellular organelles in which chemical energy is converted into a useful form (ATP). These organelles have their own genetic system.

Grobstein, C. (1979) *A Double Image of the Double Helix.* W. H. Freeman & Company, San Francisco, 177 pp.
During the mid-1970s recombinant DNA technologies led to public controversies. This book discusses the controversies and provides a copy of the NIH guidelines that regulated recombinant DNA research.

Grunstein, M. (1992) Histones as Regulators of Genes. *Scientific American,* October, pp. 68–74.
Histones are proteins that package eukaryotic DNA into ball-like structures called nucleosomes. This paper deals with the ability of histones to control gene expression.

Hakomori, S. (1986) Glycosphingolipids. *Scientific American,* May, pp. 44–53.
These fat molecules are parts of cell membranes. Cell membrane structure is described, along with changes that occur at the onset of cancer.

Hazeltine, W., and Wong-Staal, F. (1988) The Molecular Biology of the AIDS Virus. *Scientific American,* October, pp. 52–62.
One of 10 papers on AIDS in a single issue of *Scientific American.*

Hirsch, M. S., and Kaplan, J. C. (1987) Antiviral Therapy. *Scientific American,* April, pp. 76–85.
Viruses are not susceptible to the antibiotics we normally use to kill bacteria. Strategies are described for obtaining antiviral agents.

Holliday, R. (1989) A Different Kind of Inheritance. *Scientific American,* June, pp. 60–73.
The methylation of DNA may be an important way in which gene activity patterns are passed from one generation to another.

Hunter, T. (1984) The Proteins of Oncogenes. *Scientific American,* August, pp. 70–79.
Ways in which oncogenes may cause cancer.

Janeway, C. A., Jr. (1993) How the Immune System Recognizes Invaders. *Scientific American,* September, pp. 72–79.
Recombination of gene fragments generates millions of receptors needed to identify and attack pathogens.

Johnson, H. M., Bazer, F. W., Szente, B. E., and Jarpe, M. A. (1994) How Interferons Fight Disease. *Scientific American,* May, pp. 68–75.
Interferon structure and action.

Jonathon, P., Butler, G., and Klug, A. (1978) The Assembly of a Virus. *Scientific American,* November, pp. 62–69.

This description of the assembly of tobacco mosaic virus provides an introduction to the details of virus structure.

Kandel, E. R., and Hawkins, R. D. (1992) The Biological Basis of Learning and Individuality. *Scientific American,* September, pp. 78–86.
Nerve cell action and experimental efforts to understand the biochemistry of learning.

Kennedy, R. C., Melnick, J. L., and Dreesman, G. R. (1986) Anti-Idiotypes and Immunity. *Scientific American,* July, pp. 48–69.
Discussion of antibodies that recognize other antibodies, a process that is probably important in the modulation of the normal immune system.

Kornberg, R. D., and Klug, A. (1981) The Nucleosome. *Scientific American,* February, pp. 52–64.
Higher cells package their DNA by wrapping it around ball-like structures made of protein.

Lake, J. A. (1981) The Ribosome. *Scientific American,* August, pp. 84–97.
A three-dimensional model.

Laurence, J. (1985) The Immune System in AIDS. *Scientific American,* December, pp. 84–93.
A description of T4 lymphocytes, one of the cell types attacked by the AIDS virus.

Lawn, R. M., and Vehar, G. A. (1986) The Molecular Genetics of Hemophilia. *Scientific American,* March, pp. 48–65.
Hemophilia is caused by a defective gene. The product of the corresponding normal gene has now been produced artificially.

Leder, P. (1982) The Genetics of Antibody Diversity. *Scientific American,* May, pp. 102–115.
The shuffling of segments of DNA and RNA that occurs during the formation of antibody genes is considered.

Lichtenstein, L. (1993) Allergy and the Immune System. *Scientific American,* September, pp. 117–124.
In allergic individuals, parts of the immune system misdirect their power at innocuous substances.

Maniatis, T., Hardison, R. C., Lacy, E., Lauer, J., O'Connell, C., Quon, D., Sim, G., and Efstratiadis, A. (1978) The Isolation of Structural Genes from Libraries of Eucaryotic DNA. *Cell,* 15:687–701.
A research paper describing the cloning of rabbit hemoglobin genes.

Marrack, P., and Kappler, J. (1986) The T Cell and Its Receptor. *Scientific American,* February, pp. 36–45.
T cells are an important factor in the ability of the immune system to react specifically to viruses. This article introduces the many factors involved in an immune response.

Marrack, P., and Kappler, J. W. (1989) How the Immune System Recognizes the Body. *Scientific American,* September, pp. 81–89.
A description of the immune system with special reference to how it distinguishes self from non-self.

Matthews, T., and Bolognesi, D. (1988) AIDS Vaccines. *Scientific American,* October, pp. 120–127.
A good treatment of the immune response and vaccine development.

McKnight, S. L. (1991) Molecular Zippers in Gene Regulation. *Scientific American,* April, pp. 54–64.
The interaction of polypeptide subunits of a protein.

Mills, J., and Masur, H. (1990) AIDS-Related Infections. *Scientific American,* August, pp. 50–57.
Opportunistic infections are the major killers in AIDS.

Mitchison, A. (1993) Will We Survive? *Scientific American,* September, pp. 136–144.
The constant struggle between the human immune system and emerging pathogens.

Mullis, K. (1990) The Unusual Origin of the Polymerase Chain Reaction. *Scientific American,* April, pp. 56–65.
PCR is a very powerful way to amplify specific regions of a DNA molecule. This is a personal story of the discovery of PCR.

Murray, A. W., and Kirschner, M. W. (1991) What Controls the Cell Cycle? *Scientific American,* March, pp. 56–63.
A protein has been discovered that seems to be a regulator of the cell cycle in many organisms.

Murray, A. W., and Szostak, J. W. (1987) Artificial Chromosomes. *Scientific American,* November, pp. 62–87.
The behavior of whole chromosomes can be studied by creating artificial chromosomes using genetic engineering.

Nathans, J. (1989) The Genes for Color Vision. *Scientific American,* February, pp. 42–49.
A molecular explanation for color blindness is beginning to emerge.

Neufeld, P., and Coleman, N. (1990) When Science Takes the Witness Stand. *Scientific American,* May, pp. 46–53.
Some of the pros and cons of DNA fingerprinting that existed in the early days of DNA forensics.

Nomura, M. (1984) The Control of Ribosome Synthesis. *Scientific American,* January, pp. 102–115.
A description of ribosomes and the genes that encode their components.

Nossal, G. (1993) Life, Death, and the Immune System. *Scientific American,* September, pp. 52–63.
An overview of the immune system.

Paabo, S. (1993) Ancient DNA. *Scientific American*, November, pp. 86–92.
A description of work carried out on DNA extracted from long-dead plants and animals.

Paul, W. (1993) Infectious Diseases and the Immune System. *Scientific American*, September, pp. 91–97.
The focus of this paper is protection from bacterial disease.

Pestka, S. (1983) The Purification and Manufacture of Human Interferons. *Scientific American*, August, pp. 36–43.
Human cells release a protein called interferon, which protects us against infection by some viruses. Its manufacture by gene cloning techniques is described.

Prusiner, S. B. (1995) The prion diseases. *Scientific American*, January, pp. 48–51.
These disease-causing agents do not contain genetic material.

Ptashne, M. (1989) How Gene Activators Work. *Scientific American*, January, pp. 40–47.
The interaction of specific proteins with DNA can lead to transcription of certain genes.

Ptashne, M., Johnson, A. D., and Pabo, C. O. (1982) A Genetic Switch in a Bacterial Virus. *Scientific American*, November, pp. 128–140.
When bacteriophage lambda infects a cell, it can either kill its host or coexist with it. This article describes how the virus makes the choice.

Radman, M., and Wagner, R. (1988) The High Fidelity of DNA Duplication. *Scientific American*, August, pp. 40–46.
How three enzymes work together to synthesize DNA, to proofread, and to correct mistakes.

Rennie, J. (1990) The Body Against Itself. *Scientific American*, December, pp. 106–115.
The immune system is discussed with particular attention on autoimmune diseases.

Rennie, J. (1993) DNA's New Twists. *Scientific American*, March, pp. 122–132.
A historical perspective of transposons.

Rennie, J. (1994) Grading the Gene Tests. *Scientific American*, June, pp. 88–97.
Some of the ethical issues involved in genetic testing.

Richards, F. M. (1991) The Protein Folding Problem. *Scientific American*, January, pp. 54–63.
To be active, proteins must fold in a very precise way. Some ideas about folding and protein structure are discussed.

Rhodes, D., and Klug, A. (1993) Zinc Fingers. *Scientific American*, February, pp. 56–59.
Zinc fingers are protrusions from proteins that are important in allowing some proteins to interact with DNA.

Ross, J. (1989) The Turnover of Messenger RNA. *Scientific American*, April, pp. 48–55.
Genes can be controlled by way of timing the breakdown of messenger RNA.

Sachs, L. (1989) Growth, Differentiation and the Reversal of Malignancy. *Scientific American*, January, pp. 40–47.
Growth and differentiation are described with a focus on leukemia.

Sapienza, C. (1990) Parental Imprinting of Genes. *Scientific American*, October, pp. 52–60.
Identical genes can have different effects depending on whether they came from the father or the mother.

Scheller, R. H., and Axel, R. (1984) How Genes Control an Innate Behavior. *Scientific American*, March, pp. 54–83.
Egg-laying behavior in a marine snail is coupled to the expression of specific genes.

Simons, K., Garoff, H., and Helenius, A. (1982) How an Animal Virus Gets Into and Out of Its Host Cell. *Scientific American*, February, pp. 58–66.
The virus causes the host cell to manufacture new virus particles, including a protective membrane that originates from the host cell membrane.

Smith, K. (1990) Interleukin-2. *Scientific American*, March, pp. 50–57.
Interleukin-2 is thought to be a hormone that helps control the immune system.

Stahl, F. W. (1987) Genetic Recombination. *Scientific American*, February, pp. 90–101.
DNA molecules naturally break and rejoin with other DNA molecules to give new combinations of nucleotide sequences.

Steinman, L. (1993) Autoimmune Disease. *Scientific American*, September, pp. 106–114.
Several autoimmune diseases are described, along with experimental treatment.

Steitz, J. (1988) Snurps. *Scientific American*, June, pp. 56–63.
Snurps are small nuclear ribonucleoproteins involved in the removal of introns from RNA in higher cells.

Tiollais, P., and Buendia, M. A. (1991) Hepatitis B Virus. *Scientific American*, April, pp. 116–123.
This deadly virus is being attacked by recombinant DNA techniques.

Tjian, R. (1995) Molecular Machines That Control Genes. *Scientific American*, February, pp. 54–61.
A description of eukaryotic transcription factors.

Tonegawa, S. (1985) The Molecules of the Immune System. *Scientific American*, October, pp. 122–131.
A detailed description of antibodies and the genetic rearrangements that lead to antibody diversity.

Unwin, N. (1984) The Structure of Proteins in Biological Membranes. *Scientific American,* February, pp. 78–95.
The discussion includes an explanation of how protein structure is studied.

Varmus, H. (1987) Reverse Transcription. *Scientific American,* September, pp. 56–65.
The conversion of RNA to DNA is a central step in the infection of cells by retroviruses such as the AIDS virus. But this process also occurs in a number of other biological systems, and it may even reflect a very early process in the evolution of genetic material.

Verma, I. (1990) Gene Therapy. *Scientific American,* November, pp. 68–84.
A discussion of the feasibility of introducing healthy genes into people to correct diseased genes.

Wang, J. C. (1982) DNA Topoisomerases. *Scientific American,* July, pp. 94–109.
A class of enzyme is described that is able to convert rings of DNA from one topological form to another. These enzymes are probably important in many aspects of chromosome function.

Watson, J. D. (1980) *The Double Helix,* edited by Gunther Stent (1980) W. W. Norton & Company, New York, 298 pp.
Watson's account of the discovery of DNA structure is placed in historical perspective by Gunther Stent, Francis Crick, Linus Pauling, and Aaron Klug. Stent has also collected background materials, which include reproductions of original scientific papers and reviews of Watson's story by well-known scientists.

Watson, J. D., and Tooze, J. (1981) *The DNA Story.* W. H. Freeman & Company, San Francisco, 605 pp.
Included in this documentary history of gene cloning are reproductions of press clippings and personal letters written by the participants. The section titled "Scientific Background" provides a historical introduction to gene cloning.

Weinberg, R. A. (1983) A Molecular Basis of Cancer. *Scientific American,* November, pp. 126–143.
The relationship of oncogenes and cancer cells.

Weinberg, R. A. (1985) The Molecules of Life. *Scientific American,* October, pp. 48–57.
An overview of DNA, proteins, and gene cloning.

Weintraub, H. (1990) Antisense RNA and DNA. *Scientific American,* January, pp. 40–46.
Antisense oligonucleotides are capable of blocking gene expression, and they have potential as therapeutic agents.

Weissman, I., and Cooper M. (1993) How the Immune System Develops. *Scientific American,* September, pp. 64–71.

> The genetic and environmenal signals involved in differentiation of immune cells.

Welch, W. J. (1993) How Cells Respond to Stress. *Scientific American,* May, pp. 56–64.

> During emergencies cells produce stress proteins. The properties of these proteins are described.

Wigzell, H. (1993) The Immune System as a Therapeutic Agent. *Scientific American,* September, pp. 126–134.

> Focuses on vaccine development.

Winkler, W. B., and Bogel, K. (1992) Control of Rabies in Wildlife. *Scientific American,* June, pp. 86–92.

> Rabies is spreading rapidly in the eastern United States, and efforts are being made to immunize wildlife.

White, R., and Lalouel, J. (1988) Chromosome Mapping with DNA Markers. *Scientific American,* February, pp. 40–48.

> How restriction fragment length polymorphisms are used to study the genetics of disease.

GLOSSARY

adenine (A) one of the bases that forms a part of DNA or RNA; the others are cytosine, guanine, thymine, and uracil. (Figure 2-5)

agar a gelatinlike substance obtained from seaweed. When mixed with nutrients and allowed to solidify in petri dishes, agar serves as a solid substrate for growing bacterial colonies. (Figure 5-2)

agar plate a petri dish containing solid agar. (Figure 5-2)

amino acid a small molecule that serves as a subunit of protein. Twenty different types of amino acid are commonly found in natural proteins, and they share the structure shown below. The letter R represents chemical side chains, which are different for each amino acid. The chemical properties of the side chains help determine how a protein folds; thus the arrangement of amino acids dictates the three-dimensional structure of a protein.

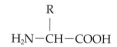

$$\overset{\overset{\textstyle R}{\textstyle |}}{H_2N-CH-COOH}$$

amino acids

alanine (Ala)	leucine (Leu)
arginine (Arg)	lysine (Lys)
asparagine (Asn)	methionine (Met)
aspartic acid (Asp)	phenylalanine (Phe)
cysteine (Cys)	proline (Pro)
glutamine (Gln)	serine (Ser)
glutamic acid (Glu)	threonine (Thr)
glycine (Gly)	tryptophan (Trp)
histidine (His)	tyrosine (Tyr)
isoleucine (Ile)	valine (Val)

aminoacyl–tRNA synthetases members of a class of enzyme that link specific amino acids with specific transfer RNA molecules. One synthetase recognizes one particular type of transfer RNA and one particular type of amino acid. (Figure 4-3)

antibiotic a substance, produced by a microorganism, that inhibits the growth of bacteria, often killing them. Most antibiotics in clinical use have been extensively modified to increase their potency. Common examples are streptomycin, erythromycin, penicillin, ampicillin, and tetracycline. (p. 93)

antibiotic resistance gene a gene encoding a protein which allows a bacterium to live in the presence of an antibiotic that normally would kill it. Some resistance genes change the target of the drug so it no longer binds the drug. Others cause active secretion of the drug, and still others break down the drug. Plasmids often contain such genes. (p. 93).

antibodies proteins found in higher animals that recognize and bind to foreign proteins such as viruses and bacteria. After binding the foreign protein (often called an antigen), the antibody can participate in a variety of reactions that lead to the destruction of the antigen. Antibodies are an important component of the immune system, which serves to guard us from attack by microorganisms. (Figure 11-7)

anticodon a particular three-nucleotide region in transfer RNA that is complementary to a specific three-nucleotide codon in messenger RNA. Alignment of codons and anticodons is the basis for establishing the order of amino acids in a protein chain. (Figure 4-4)

antigen a microorganism or foreign molecule that is recognized by, and attaches to, an antibody. (Figure 11-10)

antigenic determinant a region of an antigen that elicits an immune response.

antisense RNA an RNA molecule that is the complement of another RNA molecule and can therefore form a double helix. Antisense RNA molecules can be designed to hybridize with particular mRNA molecules and thereby prevent the translation of the mRNA.

assay a method or way of measuring chemical or biological compounds. (Figure 3-8)

atom a particle composed of a nucleus (protons and neutrons) and electrons. Atoms differ from one another by having different numbers of protons, neutrons, and electrons. Groups of atoms bonded together are called molecules. See **element.** (p. 3; Figure 1-3)

ATP adenosine triphosphate, a relatively small molecule that serves as an energy carrier and as one of the precursors to RNA. ATP has high energy bonds that are easily broken by enzymes to release the energy needed to drive many cellular chemical reactions.

B lymphocyte a type of cell in mammals that produces antibodies. (Figure 11-10)

bacterial culture a batch of bacterial cells grown in a brothlike solution (p. 87)

bacterial transformation see **transformation.**

bacteriophage a virus that attacks bacteria; also called a phage. (Figure 6-5)

bacterium (plural, bacteria) a one-celled prokaryotic organism. Although many biochemical properties of bacteria differ in detail from those of higher organisms, the basic features of chemical reactions are very similar in bacteria and man. (Figure 5-1)

base (1) a flat ring structure, containing nitrogen, carbon, oxygen, and hydrogen, that forms part of one of the nucleotide links of a nucleic acid chain. The different bases are adenine, thymine, guanine, cytosine, and uracil, commonly abbreviated A, T, G, C, and U. (2) A hydrogen ion acceptor, such as sodium hydroxide (lye). (Figure 2-2)

base pair (bp) two bases, one in each strand of a double-stranded nucleic acid molecule, that are attracted to each other by weak chemical inter-actions. Only certain pairs form: A · T (or T · A), G · C (or C · G), and A · U (or U · A). (Figures 2-3 and 2-5)

binding attaching.

broth a liquid culture medium used to grow bacteria. One common type contains yeast extract, beef extract, table salt, and water.

carcinogen a chemical that causes cancer, generally by altering the structure of DNA; see **mutagen.**

cell the smallest unit of living matter capable of self-perpetuation; an orga-nized set of chemical reactions capable of reproduction. A cell is bounded by a membrane that separates the inside of the cell from the outer environ-ment. Cells contain DNA (where information is stored), ribosomes (where proteins are made), and mechanisms for converting energy from one form to another. (pp. 2, 12)

cell extract or **lysate** a mixture of cellular components obtained by mechani-cally or enzymatically breaking cells. The cell extract is the starting mate-rial from which biochemists obtain enzymes, RNA, and DNA.

cell wall a thick, rigid structure surrounding cells of certain types, especially those of bacteria and plants. Cell walls are often composed of complex sugars.

cellulose a large carbohydrate (sugar) polymer that comprises much of the tissue of plants.

centrifuge a machine that uses centrifugal force, generated by a spinning motion, to separate molecules of various sizes and densities. Centrifuges can create forces hundreds of thousands of times that of gravity, making it possible to quickly separate molecules on the basis of size and shape. Merry-go-rounds and the spin cycle mechanisms of automatic clothes-washing machines are examples of centrifuges. (Figure 3-7)

chemical reaction a rearrangement of atoms to produce a set of molecules that are different from the starting set of molecules. (p. 7)

chloroplasts membrane-bound organelles in plants that contain chlorophyll and convert the energy of sunlight into chemical energy stored in sugars.

chromosome a subcellular structure containing a long, discrete piece of DNA plus the proteins that organize and compact the DNA. (p. 12; Figure 13-5)

cholesterol a fatty substance that is important for cellular function but can also contribute to blockage of arteries when in excess. (p. 277)

clone (1) noun: a group of identical cells, all derived from a single ancestor. (2) verb: to perform or undergo the process of creating a group of identical cells or identical DNA molecules derived from a single ancestor.

cloning vector see **cloning vehicles.**

cloning vehicles small plasmid, phage, or animal virus DNA molecules into which a DNA fragment is inserted to enable the transfer of the fragment from a test tube into a living cell. Cloning vehicles are capable of multiplying inside living cells. Thus, if a cloning vehicle transfers a specific fragment of DNA into a cell that is also multiplying, all the progeny of that cell will contain identical copies of the vehicle and the transferred DNA fragment. (Figures 6-1, 6-5)

code (genetic) the system in which the arrangement of nucleotides in DNA represents the arrangement of amino acids in protein. (Figure 2-6)

codon a sequence of three adjacent nucleotides whose precise order and location in a nucleic acid specifies the insertion of a particular amino acid into a specific structural position of a protein during protein synthesis. Special codons that do not code for any amino acid act as stop signals. In some cases several different codons encode the same amino acid. (Figure 2-6)

cointegrate a DNA molecule derived from the joining of two dissimilar plasmid DNA molecules. (Figure 11-14)

colony a visible cluster of microorganisms on a solid surface. All members of the colony arise from a single parental cell, and the colony is considered to be a clone. In the case of bacteria, a colony generally contains millions of individual cells. (Figure 5-2)

column chromatography a process by which different molecule types are separated on the basis of the speed at which they pass through a glass cylinder filled with one of many types of solid materials to which the molecule types bind with different affinities. (Figure 3-9)

competence, bacterial a state in which bacterial cells readily take up DNA molecules from the environment. (p. 103)

complementary describing two objects having shapes that allow them to fit together very closely: plugs and sockets, locks and keys, molecules of G and C, molecules of A and T or U.

complementary base-pairing rule the principle according to which only certain nucleotides can align opposite each other in the two strands of a double helical nucleic acid: G pairs with C; A pairs with T (or U in RNA). (Figure 2-5)

complementary DNA (cDNA) DNA synthesized from RNA in test tubes using an enzyme called reverse transcriptase. The DNA sequence is thus complementary to that of the RNA. Complementary DNA is often prepared with radioactive nucleotides and is used as a hybridization probe to detect specific RNA or DNA molecules. When cDNA is prepared from cytoplasmic mRNA of a eukaryotic organism, it represents the nucleotide sequences responsible for encoding proteins. Total DNA includes these sequences plus those that are removed from mRNA by RNA splicing and those that are not transcribed into RNA.

conjugation, bacterial the process whereby two bacterial cells join together and DNA is passed from one to the other. (Figure 6-2)

constant region a portion of an antibody molecule that is the same within a particular class of antibody. (Figure 11-7)

contig a long contiguous region of DNA in which all of the sections have been cloned.

cystic fibrosis a recessive genetic disease caused by the inability of cells to properly secrete salt. Among the symptoms is formation of thick mucus in the lungs.

cytological relating to the study of cell structure, often using microscopy. A commercial cytology laboratory examines the structure of chromosomes for genetic disorders.

cytoplasm the interior portion of a cell exclusive of the nucleus.

cytosine (C) one of the bases that forms a part of DNA or RNA. (Figure 2-5)

denatured unfolded; rendered inactive. In reference to DNA, denaturation means conversion of double-stranded DNA into single-stranded DNA. In reference to proteins, denaturation means unfolding of the protein.

density gradient a solution, usually of salt or sugar, in which the density gradually increases from top to bottom as a result of the increasing concentration of salt or sugar. (Figure 6-4)

deoxyribonucleic acid see **DNA.** (Figure 2-3)

dissolve to disperse a solid substrate in a liquid.

DNA deoxyribonucleic acid. DNA is a long, thin chainlike molecule that is usually found as two complementary chains and is often hundreds to thousands of times longer than the cell in which it resides (it is tightly wrapped to fit inside). The links or subunits of DNA are the four nucleotides called deoxyadenylate, deoxycytidylate, deoxythymidylate, and deoxyguanylate. The precise arrangement of these four subunits is used to store all information necessary for life processes. (Figures 1-1, 2-3)

DNA fingerprinting the use of several regions of DNA that vary from person to person to identify the source of a tissue sample. (p. 281)

DNA footprinting an experimental procedure in which the sites of protein binding to DNA are determined by treating protein-DNA complexes with agents that damage DNA regions not protected by the bound protein. After removal of the protein the damaged and undamaged regions are identified with respect to nucleotide sequence. (Figure 10-8)

DNA ligase the enzyme that joins two separate DNA molecules together end to end.

DNA polymerase the enzyme complex that makes new DNA using the nucleotide sequence information contained in old DNA. (Figure 3-3)

DNA replication the process of making DNA. DNA is always made from preexisting information in DNA (or, in special cases, from RNA). DNA replication involves a number of different enzymes. (Figure 3-1)

dominant a form of a gene that specifies a characteristic of an organism regardless of whether present in both or only in one of the two chromosomal copies. (Figure 13-2)

E. coli *Escherichia coli,* a species of bacterium commonly found in mammalian digestive tracts. (Figure 5-1)

egg (1) ovum or germ cell produced by a female; (2) an animal embryo, along with a food supply, enclosed by a shell or membrane.

electron micrograph a photograph taken using an electron microscope. (Figures 6-1, 6-5)

electron microscope an instrument that is similar to a light microscope but uses a beam of electrons to expose film rather than a beam of light. Because the effective wavelength of electrons is much shorter than that of light, much smaller objects can be seen with an electron microscope than with a light microscope. Objects that are measured in millionths of a centimeter can be seen using an electron microscope.

electrophoresis a process in which an electric field is used to separate molecules that differ in charge. (Figure 7-4)

element one of slightly more than 100 distinct types of matter, which singly and in combination compose all materials of the universe. An atom is the smallest representative unit of an element. Common elements are oxygen, hydrogen, carbon, and nitrogen. (Figure 1-3)

embryo a plant or an animal at an early stage of development, generally before it has attained a form easily recognized as a distinct species of organism.

encode to contain a nucleotide sequence specifying that one or more specific amino acids be incorporated into a protein. (Figure 1-6)

endonuclease an enzyme that cuts DNA or RNA at points inside the nucleic acid molecule (i.e., away from the ends). (Figure 7-1)

enhancer a region of DNA that stimulates initiation of transcription or replication even though it is far from the promoter or origin of replication. (Figure 4-9)

envelope a covering or coat; the outermost coat of an animal virus. (p. 229; Figure 12-1)

enzyme a protein molecule, or occasionally an RNA molecule, specialized to catalyze (accelerate) a biological chemical reaction without itself being consumed. Generally enzyme names end in -ase. (p. 7; Figure 4-3)

equilibrium the absence of *net movement* one way or another.

eukaryote any organism in which the genetic material is localized in a membrane-bound organelle called a nucleus.

exon a region of RNA that encodes a portion of a protein; exons remain in the mRNA after introns have been removed by splicing. (Figures 11-4, 11-5)

expression see **gene expression.**

expression vector a plasmid or phage in which the DNA contains an active, easily controlled promoter, downstream from which a gene of interest can be inserted. Following induction of the promoter, the protein of interest can be produced in large amounts, sometimes up to 40% of the total protein of the cell that carries the vector. (Figure 10-7)

extract (1) verb: to separate one type of molecule from all others, to purify; (2) noun: a mixture of molecules obtained by breaking cells.

fertility factor, F a large plasmid that carries genes necessary for bacterial conjugation. (Figure 6-2)

fetus an embryo in a late stage of development, but still in the uterus; the adjectival form is **fetal.**

fission a type of cell division in which a parental cell divides in half to form two daughter cells. (Figure 3-2)

five-prime (5') and three-prime (3') ends the backbone of a nucleic acid molecule is composed of repeating phosphate and sugar subunits. The carbon atoms in the sugar are numbered 1' through 5'. On one side of the sugar, the phosphate is linked to the 5' carbon of the sugar, and on the other side the phosphate is linked to the 3' carbon of the sugar. When a chain is broken, the break generally occurs between the phosphate and the sugar. This produces two different ends. If only the sugar is considered, a 5' carbon will be at one end (the 5' end) and a 3' carbon will be at the other (the 3' end). These terminal carbons generally have a phosphate (PO_4) or a hydroxyl (OH) group attached to them. (Figure 2-4)

foci clumps of animal cells that grow into small piles on top of a monolayer of cells. The clumps arise because the cells have been transformed into tumor cells by virus infection. Unlike normal cells, transformed cells do not stop proliferating once a monolayer forms in the laboratory dish where they are growing.

frameshift mutation an insertion or removal of one or more nucleotides that changes the reference frame with which codons are defined, thus causing the placement of incorrect amino acids in a protein from the point of the mutation to the end of the coding sequence. (Figure 3-6)

gel electrophoresis a method for separating molecules based on their size and electric charge. Molecules are driven through a gel (e.g., gelatin) by placing them in an electric field. The speed at which they move depends on their size and charge. (Figures 7-4, 10-6)

gene a small section of DNA that contains information for construction of one protein molecule, or in special cases for construction of transfer RNA or ribosomal RNA. (Figures 1-2, 4-7)

gene cloning a way to use microorganisms to produce millions of identical copies of a specific region of DNA. (p. 13; Figure 9-1)

gene expression the process of making the product of a gene; information is transferred, via messenger RNA, from a gene to ribosomes, where a specific protein is produced. (Chapter 4)

gene therapy a process of altering genes in living persons through the administration of specific regions of DNA. (p. 278)

genetic engineering the manipulation of the information content of an organism to alter the characteristics of that organism. Genetic engineering may use simple methods such as selective breeding or complicated ones such as gene cloning.

genetic map a representation of DNA in which the relative position of regions is determined by the frequency of genetic recombination between observable traits. (Figure 4-7)

genome the genetic information of an organism or virus; for organisms with two pairs of each chromosome, the genome refers to the information in one set. The genome of a eukaryotic cell does not include the genetic information in organelles such as mitochondria and chloroplasts, which have their own genomes.

genomic clone a clone obtained from DNA representing the entire genome as opposed to a cDNA clone, which is obtained from mRNA.

germ cell a particular type of cell (sperm or egg) responsible for creating the next generation; also called a gamete. In most higher organisms body cells contain two sets of chromosomes; germ cells contain only one set. Thus when two germ cells fuse, the resulting cell (zygote) has two sets of chromosomes. This cell then produces new body (somatic) cells.

germ line gene therapy the process of altering genes in germ cells, usually by means of recombinant DNA technology, such that the altered genes are passed from one generation to the next. (p. 280)

globins a family of proteins involved in movement of oxygen in blood and tissues. (Figures 11-2, 11-3)

glycosylation the process of adding long sugar chains to protein molecules, often to direct the proteins to specific cellular compartments or to specific body tissues. Glycosylation may also serve to protect the protein from attack by proteases.

guanine (G) one of the bases that form a part of DNA or RNA. (Figure 2-5)

heavy chain the larger of the protein molecules that come together to form an antibody (antibodies are composed of two heavy chains and two light chains). (Figure 11-7)

hemoglobin an iron-containing protein involved in the movement of oxygen in the bloodstream. (Figure 11-2)

heredity the genetic transfer of characteristics from parents to offspring. (Figure 13-4)

heteroduplex; heteroduplex mapping double-stranded DNA in which one strand is from one source and the second strand is from another source such that the strands are complementary only over part of their lengths. Complementary and noncomplementary regions can be distinguished by electron microscopy, making it possible to construct physical maps of DNA. (Figure 6-1b)

histones members of a small class of eukaryotic chromosomal proteins that wrap DNA into ball-like structures called nucleosomes.

homologous corresponding or similar in position; describing regions of DNA molecules that have the same nucleotide sequence. Since DNA has two complementary strands, complementary base pairing can occur between homologous regions in two different DNA molecules. Regions of DNA, RNA, or protein that are similar due to a common ancestry also are said to be homologous.

homologous chromosomes a pair of chromosomes, one derived from the father and one from the mother, that contain the same set of genes. (Figure 13-5)

host an organism that provides the life support system for another organism, virus, or plasmid. *E. coli* is a host for certain plasmids that exist inside the bacterium, and we are hosts for *E. coli*, since these bacteria live inside us and many other mammals.

human genome the information content of one set of human chromosomes.

hybrid, nucleic acid a double-stranded nucleic acid in which the two complementary strands differ in origin. One strand can be RNA and the other DNA, or both strands can be either RNA or DNA. (Figure 8-1)

hybridoma a hybrid cell type formed from the fusion of a B lymphocyte with a cancerous antibody-producing cell called a myeloma. (p. 212)

hydrogen bond a weak attraction between a hydrogen atom and usually oxygen or nitrogen. Hydrogen bonds are important forces that stabilize DNA double helices and facilitate the formation of complementary base pairs. (Figure 2-5)

induce to cause to happen, often with reference to gene expression. Specific molecules called inducers bind to particular repressors and prevent the repressor from binding to DNA. That allows gene expression to occur.

infectious capable of invading a host. (p. 98)

insulin a protein involved in the control of sugar metabolism in mammals. Insulin is made by cells of the pancreas.

integrase a protein involved in the insertion of one DNA molecule into another. (Figure 12-6)

integration the process of inserting one DNA molecule into another. (Figures 6-8, 12-6)

intron a region of RNA that is removed by splicing and thus does not contribute information for the formation of protein. (Figure 11-4)

kinase an enzyme that adds a phosphate to another molecule.

lambda a type of bacteriophage that can either lyse its bacterial host or take up residence in the host and allow the host to live. (Figure 6-5)

leader a region of film, RNA, or protein that precedes the region of primary information content. In RNA, the leader extends from the first nucleotide at the 5' end to the codon specifying the first amino acid in the protein. A protein leader, if present, is usually defined as a region at the amino terminal end that is cut from the protein during movement of the protein across membranes. (Figure 4-8)

light chain the smaller of the proteins that make up an antibody. (Figure 11-7)

light microscope an optical instrument that uses light and a combination of lenses to produce magnified images of small objects.

long terminal repeat (LTR) long terminal repeat nucleotide sequence found at each end of a retroviral genome in the DNA form. (Figure 12-3)

lysogeny the process by which a bacteriophage enters a dormant state during infection of a bacterial cell. The bacteriophage can later be stimulated to kill the cell. (Figure 6-8)

lysosome a membrane-bound organelle of eukaryotic cells that contains degradative enzymes; equivalent to a cellular garbage disposal.

lytic infection a type of viral infection in which the host cells are broken and virus is released.

macromolecule a very large molecule, usually composed of many smaller molecules joined together.

meiosis the type of cell division that occurs during formation of reproductive cells to reduce the number of chromosomes from two to one of each type.

membrane a thin sheet; for cells, membranes surround collections of molecules to provide distinct compartments. This includes a membrane around the entire cell. Biological membranes are generally composed of two layers of **phospholipids** plus assorted proteins that act as channels for movement of molecules and as receptors for binding of signals from outside the cell. (Figure 12-1)

messenger RNA (mRNA) RNA used to transmit information from a gene on DNA to a ribosome, where the information is used to make protein. (Figure 4-4)

metabolism a collective term for the chemical reactions involved in life. For example, sugar metabolism includes the reactions that occur in the body during the production, use, and breakdown of sugars.

micrometer (μm) one millionth of a meter (1 meter = about 39 inches).

milligram (mg) one thousandth of a gram (28 grams = 1 ounce).

millimeter (mm) one thousandth of a meter (1 meter = about 39 inches).

mitochondrion (plural, mitochondria) a specialized intracellular structure (organelle) that converts the chemical energy stored in food into a more useful form as molecules called ATP.

molecule a group of atoms tightly joined together. The arrangement of atoms is very specific for a given molecule, and this arrangement gives each molecule specific chemical and physical properties. Each molecule of oxygen we breathe is two oxygen atoms bonded together. Paper consists largely of cellulose molecules, which are giant molecules containing carbon, oxygen, and hydrogen. (Figures 1-1, 1-4)

monoclonal derived from a single clone of cells, often in reference to an antibody. In a population of monoclonal antibodies, all members are identical. (p. 212)

monolayer a layer of cells one cell thick.

mutagen an agent that increases the rate of mutation by causing changes in the nucleotide sequences of DNA; see **carcinogen.**

mutant an organism whose DNA has been changed relative to the DNA of the dominant members of the population. (Figure 3-6)

mutations errors in DNA, often occurring during DNA replication, that cause proteins to have an incorrect amino acid sequence. (Figure 3-6)

myeloma cells cancerous cells that produce antibodies. (p. 212)

N-terminus the end of a protein chain that is bounded by an NH_2 (amino) group. The N-terminus is the end of a protein that is synthesized first.

Northern blotting a form of transfer hybridization in which the target nucleic acid is RNA. Also called Northern transfer hybridization.

nuclease a general term for an enzyme that cuts DNA or RNA.

nucleic acid DNA, RNA, or a DNA : RNA hybrid.

nucleic acid hybridization a process in which two single-stranded nucleic acids are allowed to base-pair and form a double helix. The process makes it possible to use one nucleic acid to detect the presence of another having a complementary nucleotide sequence. (Figure 8-1)

nucleoid intact, compact bacterial chromosome either inside a cell or extracted by gentle methods that do not break DNA (Figures 1-1, 5-1b)

nucleotide one of the building blocks of nucleic acids. A nucleotide is composed of three parts: a base, a sugar, and a phosphate. The sugar and the phosphate form the backbone of the nucleic acid, while the bases lie flat like steps of a spiral staircase. DNA is composed of deoxyadenylate, deoxythymidylate, deoxyguanylate, and deoxycytidylate, four different types of nucleotide represented by the letters dA, dT, dG, and dC (abbreviations A, T, G, and C are sometimes used). See also **RNA** and **sequence.** (Figure 2-2)

nucleotide pair two nucleotides, one in each strand of a double-stranded nucleic acid molecule, that are attracted to each other by weak chemical interactions between the bases. Only certain pairs, A · T, G · C, and A · U, form with high affinity. (Figure 2-3)

nucleoside triphosphate a molecule composed of a base, sugar, and three phosphates in the order base-sugar-phosphate-phosphate-phosphate. During synthesis of DNA and RNA the two end phosphates are cut off to provide energy for joining of the inner phosphate to another sugar.

nucleus (1) the core of an atom consisting of protons and neutrons; (2) a distinct subcellular structure containing chromosomes and surrounded by a membrane. (Figure 12-4)

oligonucleotide a short piece of DNA or RNA containing three or more nucleotides. The oligonucleotides used in gene cloning are generally less than 100 nucleotides long, but in formal terms an oligonucleotide can be much longer.

oncogene a gene whose protein product promotes the loss of cellular control over division and leads to malignant growth. When inserted into normal cells, oncogenes can transform these units into tumor cells. (p. 243)

open reading frame a stretch of nucleotides, usually more than 200 nucleotides long, that is bounded by translation start and stop codons. The codons are in a reading frame that could be translated into protein.

operator a region of DNA capable of interacting with a repressor, thereby controlling the functioning of an adjacent gene. (Figure 4-6)

operon a series of bacterial genes transcribed into a single RNA molecule. Operons allow coordinated control of a number of genes whose products have related functions. (Figure 4-7)

organelle a microscopic, membrane-enclosed body found inside eukaryotic cells. Examples are nuclei, mitochondria, and chloroplasts.

organism one or more cells organized in such a way that the unit is capable of reproduction.

origin of replication a special nucleotide sequence that serves as a start signal for DNA replication. (p. 46)

pathogen a disease-causing agent (e.g., viruses that cause polio, mumps, or measles; bacteria that cause cholera, tuberculosis, or leprosy).

PCR see **polymerase chain reaction.** (Figure 8-5)

penicillin an antibiotic that kills *E. coli* and many other bacteria by blocking formation of new cell walls. Natural penicillin is produced by a mold.

peptide a short chain of amino acids, a fragment of a protein. (Figure 1-5)

peptide bond the type of chemical bond that links two adjacent amino acids together in a protein chain. (Figure 1-5)

phage a virus that attacks bacteria; abbreviation of **bacteriophage.** (Figure 6-5)

phage plaques clear zones, created by bacteriophages killing bacteria in a lawn of bacteria on an agar plate. (Figure 6-6)

phosphate a chemical unit in which four oxygen atoms are joined to one phosphorus atom. The backbones of DNA and RNA are alternating phosphate and sugar units. (Figure 2-4)

phospholipid a fatty substance (lipid) with a phosphate at one end.

physical map a representation of a DNA molecule in which the relative positions of regions are determined by physical measurements, such as by electron microscopy, restriction analysis, or sequence determination, as opposed to frequency of genetic recombination. (p. 133)

pilus (plural, pili) a filamentous appendage of bacterial cells involved in conjugation. (Figure 6-2)

pipette a precisely marked glass tube used for measuring liquid volumes.

plasmids small, circular DNA molecules found inside bacterial cells. Plasmids reproduce every time the host bacterial cell reproduces. (Figure 6-1)

plaque a hole or clear zone formed in a lawn of bacteria on an agar plate through the lethal action of bacterial viruses. (Figure 6-6)

polyclonal derived from a variety of cell lines; generally refers to a population of antibodies generated by cells that each produce a slightly different antibody molecule.

polymerase chain reaction (PCR) a test tube reaction in which a specific region of DNA is amplified many times by repeated synthesis of DNA using DNA polymerase and specific primers to define the ends of the amplified region. (Figure 8-5)

polyprotein a long protein that is cleaved into several smaller proteins. The smaller proteins are thought to be the functional forms. (p. 227)

precipitate molecules that are clumped together so that they fail to pass through a filter. Precipitates are large aggregates and settle out of solution rapidly, much like silt out of river water. (Figure 3-8)

prenatal before birth.

primer a piece of DNA or RNA that provides an end to which DNA polymerase can add nucleotides. (Figures 3-8, 8-5)

probe a DNA or RNA molecule, usually radioactive, that is used to locate a complementary RNA or DNA by hybridizing to it. Often a probe is used to identify bacterial colonies that contain cloned genes and to detect specific nucleic acids following separation by gel electrophoresis. (Figure 8-2)

product the new molecules produced by a chemical reaction.

progeny offspring.

prokaryote an organism lacking a true nucleus or other organelles.

promoter a short nucleotide sequence on DNA where RNA polymerase binds and begins transcription. (Figure 4-6)

protease an enzymatic protein that breaks down other proteins.

protein a class of long, chainlike molecules often containing hundreds of links called amino acids. Twenty different amino acids are used to make proteins. The thousands of different proteins serve many functions in the cell. As enzymes, they control the rate of chemical reactions, and as structural elements they provide the cell with its shape. Proteins are also involved in cell movement and in the formation of cell walls, membranes, and protective shells. Some proteins also help package DNA molecules into chromosomes. (Figure 1-5)

protein synthesis formation of protein; see **translation.** (Figures 4-4, 4-5)

proto-oncogene a gene that can be converted to an oncogene, pushing a cell toward a cancerous state.

provirus viral DNA in a quiescent form inside a host cell, sometimes but not always integrated into a host chromosome. With proper stimulation, the viral DNA can be induced to produce virus particles. (Figures 6-8, 12-4)

pseudogenes genes that contain many alterations, such as misplaced stop codons and frameshift mutations, that render the genetic material nonfunctional. (Figure 11-3)

purification the process of separating or isolating one type of molecule away from other types. (Figures 3-7, 3-9)

radioactive the state in which a substance (a molecule in the context of this book) contains an unstable element that spontaneously emits a high energy particle or radiation. The emission is detectable by photographic film, by Geiger counter, and by other instruments. Gene cloners generally use radioactive hydrogen, carbon, sulfur, or phosphorus, each of which is commercially available. Radioactive uranium and plutonium are used in nuclear reactors.

reading frame consecutive blocks of nucleotide triplets considered as co-dons specifying amino acids. Since nucleotide sequences can, in principle, encode three different amino acid sequences, the particular codon speci-fied by a nucleotide sequence is defined by the precise position of the translation start site. (p. 32)

recessive a form of a gene that specifies a characteristic of an organism only when present in both of the chromosome copes of the gene. (Figure 13-3)

recognition site a short series of nucleotides specifically recognized by a protein, usually leading to the binding of that protein to the DNA at or near the point of the recognition sequence. Once the protein has bound to the DNA, it may cut, modify, bend, or cover the DNA, depending on the function of the protein. (Figure 7-1)

recombinant DNA molecule a DNA molecule containing two or more regions of different origin (e.g., plasmid DNA joined to a fragment of human DNA). (Figure 7-3)

recombination the breaking and rejoining of DNA strands to produce new combinations of DNA molecules. Recombination is a natural process that generates genetic diversity. Specific proteins are involved in recombina-tion. (Figure 14-1)

replica plating the process whereby the colonies on one agar plate are transferred to another in the same relative orientation. Generally a piece of velvet (cloth) is placed on the first plate, removed, and then touched to the second plate.

replication fork the point at which the two parental DNA strands separate during DNA replication. (Figure 3-3)

replicon a DNA molecule containing the signals necessary for replication.

repression a method for preventing gene expression in which a protein molecule (repressor) binds to the DNA near where RNA polymerase ordinarily would bind. (Figure 4-6)

repressor a protein molecule that is capable of preventing transcription of a gene by binding to DNA in or near the gene. (Figure 4-6)

restriction endonucleases enzymes that cut DNA at specific nucleotide sequences. The function of this class of enzyme inside cells is to protect the cells against invasion by foreign DNA. Biologists use these enzymes as scissors to cut DNA in specific places. (p. 123)

restriction fragment length polymorphism (RFLP) a region of DNA that has several forms (the region differs from one individual to another) such that cleavage of DNA using a restriction endonuclease generates a DNA fragment whose size varies from one individual to another.

restriction mapping a procedure that uses restriction endonucleases to produce specific cuts in DNA. The positions of the cuts can be measured and oriented relative to each other to form a crude map. (p. 129)

restriction site a short, specific nucleotide sequence at which a particular restriction endonuclease cuts DNA.

retrovirus a type of animal virus whose life cycle involves conversion of genetic information from an RNA form to a DNA form. (Figure 12-1)

reverse transcriptase an enzyme purified from RNA tumor viruses that makes DNA from RNA. (Figure 12-5)

RFLP see **restriction fragment length polymorphism.**

ribonuclease H an enzyme that digests RNA but only when the RNA is hybridized to DNA. (Figure 12-5)

ribonucleic acid see **RNA.**

ribosomes large, ball-like structures that act as workbenches where proteins are made. A bacterial ribosome consists of two balls, a small one called 30S and a larger one called 50S (30S and 50S refer to speeds at which the particles sediment during centrifugation). Ribosomes are composed of special RNA molecules (ribosomal RNA) and about 50 specific proteins (ribosomal proteins). (Figure 4-2)

ribozyme an RNA molecule that acts catalytically to cleave itself or another RNA molecule. Some ribozymes have the ability to join two RNA molecules together end to end. (Figure 11-6)

RNA long, thin chainlike molecules in which the links or subunits are the four nucleotides adenylate, cytidylate, uridylate, and guanylate (abbreviated with the letters A, C, U, and G). The precise arrangement of these four subunits is used to transfer, and sometimes store, genetic information. Some RNA molecules serve as structural parts of cellular components (ribosomes), and some (transfer RNA) help align amino acids in the correct order when proteins are being made. Some RNA molecules (ribozymes) have enzymatic activity and serve as catalysts to accelerate specific chemical reactions.

RNA:DNA hybrid a double-stranded molecule composed of one strand of RNA and one of DNA. The nucleotide sequences in the DNA and RNA are complementary. (Figure 8-1)

RNA polymerase the enzyme complex responsible for making RNA from DNA. RNA polymerase binds at specific nucleotide sequences (promoters) in front of genes in DNA. It then moves through a gene and makes an RNA molecule that contains the information contained in the gene. Bacterial RNA polymerase makes RNA at a rate of about 65 nucleotides per second. (Figure 4-2)

RNA splicing the process of removing regions from RNA. The removed regions are called introns, and the regions spliced together are called exons. (Figure 11-5)

RNA tumor virus a type of RNA-containing virus that produces tumors in animals or converts normal cells in culture into tumor cells. (Figure 12-1)

sequence the order of; in reference to DNA or RNA, the order of nucleotides. (Figure 10-6)

sigma factor a subunit of RNA polymerase that allows the polymerase to recognize a specific set of promoters. (p. 80)

signal peptide a short stretch of amino acids at the N-terminal end of some proteins that direct the protein to pass through membranes. The signal peptide is often cut off as the protein passes through the membrane. (p. 70)

sodium hydroxide (NaOH) lye, caustic soda; used to denature DNA.

somatic pertaining to the body. When referring to a type of cell, somatic means body cell rather than a germ (sperm- or egg-producing) cell. Somatic cells contain two pairs of each chromosome, while germ cells contain only one.

Southern blotting (Southern hybridization) a method for transferring DNA from an agarose or acrylamide gel to nitrocellulose paper on nylon membrane followed by hybridization to a radioactive probe. Transfer hybridization using DNA as the target of a radioactive nucleic acid probe. (Figures 8-2, 8-3)

sperm germ cell produced by a male.

stem cell a cell type that has not specialized to carry out particular functions and retains the ability to divide and differentiate to form a variety of cell types.

sterile without life; generally referring to an instrument or a solution that has been heated to kill organisms that may have been on or in it. Wire is sterilized by heating in a flame until it is red hot. Culture medium (e.g., broth) is sterilized by heating in a pressure cooker (autoclave). Sterile also means unable to reproduce.

sticky ends ends of DNA that are single-stranded and complementary. (Figure 7-1)

submicroscopic too small to be seen with a light microscope.

substrate the molecules on which an enzyme acts. (Figure 3-8)

subunit one of the pieces that forms a part of a multicomponent structure, such as a link in a chain, an amino acid in a protein, or a nucleotide in DNA. (Figures 1-2, 1-5)

sugar a class of molecule containing particular combinations of carbon, hydrogen, and oxygen. The sugars in DNA and RNA are five-carbon sugars called deoxyribose and ribose, respectively. Glucose, a major constitutent of honey, is a sugar containing six atoms of carbon per molecule. (Figure 2-2)

sugar metabolism a group of biochemical reactions responsible for the formation of sugars and the conversion of sugars into other compounds.

terminus of replication a site on DNA where replication forks stop.

tetracycline an antibiotic that kills bacteria by blocking protein synthesis.

tetramers four subunits, often identical. Many proteins are composed of separate polypeptide chains that act as subunits, associating as a tetramer to form the active protein. (Figure 11-1)

three-prime (3') ends see **five-prime (5') and three-prime (3') ends.** (Figure 2-4)

thymine (T) one of the bases that forms part of DNA. It is not found in RNA. (Figure 2-2)

topoisomerase an enzyme that breaks and rejoins DNA strands in a way that changes the number of times one strand crosses the other. Topoisomerases can tie and untie DNA knots, introduce and remove DNA twists, and link and unlink DNA circles. (Figure 2-7)

toxin a substance, often a protein in the context of this book, that causes damage to the cells of an organism.

transcription the process of converting information in DNA into information in RNA. Transcription involves making an RNA molecule using the information encoded in the DNA. RNA polymerase is the enzyme that executes this conversion of information. (Figure 8-1)

transfer RNAs (tRNAs) small RNA molecules (each about 80 nucleotides long) that serve as adapters to position amino acids in the correct order during protein synthesis. The ordering by tRNA uses information in messenger RNA and occurs before the amino acids are linked together. (Figures 4-3, 4-4)

transformation the process whereby a bacterial cell takes up free DNA, with the result that information in the free DNA becomes a permanent part of the bacterial cell. Often this means introducing a plasmid into a bacterial cell. With animal cells, transformation means the conversion of a normal cell into a tumor cell.

translation the process of converting the information in messenger RNA into protein. Also called protein synthesis. (Figure 4-5)

transposase a protein encoded by a gene in a transposon that is required for transposition. (Figure 11-14)

transposition the process whereby one region of DNA moves to another. Transposition often involves duplication of the region that moves. (p. 214)

transposon a short section of DNA capable of moving to another DNA molecule or to another region of the same DNA molecule. (Figure 11-13)

tryptophan one of the 20 natural amino acids.

tumor an abnormal cluster or growth of cells that have escaped the limits on cell proliferation found in healthy organisms. (Figure 11-16)

tumor suppressor gene a gene that ordinarily inhibits the formation of tumors; suppressor genes in general reduce the effect of another gene or process. (p. 242)

turbidity cloudiness of a bacterial culture that is related to the concentration of cells.

ultraviolet light a type of light that has very high energy and is invisible; black light. Nucleic acids absorb ultraviolet light, and instruments are available that measure the amount of absorption. By measuring the amount of absorption, it is possible to determine the amount of nucleic acid present.

uracil (U) one of the bases that forms part of RNA. It is generally not found in DNA.

variable region a section of an antibody protein that varies in amino acid sequence from one antibody molecule to another. (Figure 11-7)

vector see **expression vector** and **cloning vehicles.**

virus a small pathogen that reproduces only in living cells by a process other than direct division. Viruses are composed of an RNA or DNA genome that is protected by a protein coat. In some cases enzymes are present in virus particles, and some virus coats contain lipids (fats). (Figures 6-5, 12-1)

VNTR a particular type of restriction fragment length polymorphism used for identifying organisms, especially humans. The letters represent variable number of tandem repeats." (Figure 14-5)

Walking along DNA a method for cloning DNA regions that are adjacent to ones that are already cloned. (Figure 8-4)

X-chromosome inactivation the process by which one of the two X chromosomes of a female mammal is rendered unable to express a major portion of its genes. (p. 260)

yeast a one-celled eukaryotic organism commonly used in brewing and baking. Yeasts contain mitochondria and a true nucleus. Many biochemical properties of yeast are more like those of mammals than those of bacteria. (Figure 5-3)

zygote a fertilized egg.

INDEX